VOLUME 56 AUGUST 2000 PAGES 181-372

# THE ANTECEDENTS OF NAZISM: WEIMAR

## *The Political Papers of Walter Landauer*

BY
HUGH CLARK
JULIUS ELIAS
PETER BERGMAN



*TRANSACTIONS*
THE CONNECTICUT ACADEMY OF ARTS AND SCIENCES

P.O. Box 208211
New Haven, Connecticut 06520-8211

CAAS@YALE.EDU

# TABLE OF CONTENTS

Walter Landauer (1896-1978)

Credit: Fabian Bachrach

# PREFACE

Walter Landauer's political papers represent a substantial contribution to the literature of the Weimar period, a tumultuous time. Viewing the confusion among the contenders in the aftermath of the War, and especially apprehensive about the right wing, he wrote in June, 1919: "With unchallengeable certainty, with incredibly insolent behavior, the *counter-revolution* is rising in Germany and is arming itself for a powerful assault" (paper #7), a remarkable anticipation of future events.

Having completed two years of military service he committed himself at age 22 to a rigorous curriculum in the new field of genetics at the University of Heidelberg. But he found time not only to observe the political and economic confusion but also to record his opinions about the affairs of the new government and the shortcomings of the German people. He wrote these opinions regularly for a decade, 1919 to 1929, most while he was a student, some after he came to Connecticut in 1924.

The papers, published originally in Swiss, German and Austrian newspapers and journals, have been translated by one of us (J.E.) to make them accessible to English-speaking students and general readers. Copies of the originals, long out of print, and the translations, are deposited in the Walter Landauer Collection, Archives and Special Collections, Thomas J. Dodd Research Center, University of Connecticut Libraries. Those which we believe present Landauer's principles and perspicuity most effectively are reproduced here in full. All (131) are listed in a composite index. We provide also a bibliography for further exploration of topics germane to the Landauer papers, and, when possible, explanatory or clarifying notes.

The setting: the war had ended, the monarchy had been replaced by the Weimar Republic, the National Assembly had been elected but had not yet approved a constitution. The reconstruction of Germany, harnessing its resources for peace time productivity, acceptance of the harsh terms, especially reparation payments, of the Versailles Treaty together with the inevitable political adjustments, created an atmosphere of chaos. Landauer, recognizing the vulnerability of the hard-pressed nation, was fully aware of the magnitude and diversity of Germany's troubles.

From the beginning he was a strong advocate of socialism and socialist cooperative settlements, supported in both objectives by his uncle, Gustav Landauer. Gustav was active in many ventures, including the short-lived revolutionary government in Bavaria of which indeed, he served briefly as Minister of Education. For this association he was imprisoned and brutally murdered by the *Freikorps*. That traumatic experience undoubtedly sharpened

Walter's views of the danger to Germany from what can now be seen as the rise of Nazism, but it did not blind him to the dangers of Communism. He was uncommonly prescient about the splintering of the center-left that made the Republic so vulnerable to the many faces of the revanchist right.

Storrs CT, June 2000

Hugh Clark
Julius Elias
Peter Bergmann

## ACKNOWLEDGEMENTS

The editors wish to express appreciation to the University of Connecticut Research Foundation for financial support for this and related Landauer studies. We are grateful for the technical assistance of Arlene Jacobson and Patricia Chaves in the preparation of the manuscript; Lynn Davis of the University of Hartford for scrutiny of the final text; to Ellen Clark for patient review of the introductory chapters' anecdotal items; to James A. Slater and other colleagues for helpful discussion of Landauer's views; and to Thomas F. Malone for critical comments on the moral and historical relevance of the Landauer Collection. Betsy Pittman, archivist in the Thomas Dodd Research Center, has cooperated in many ways related to the Landauer archive. Professor Karl Holl, University of Bremen, transferred to us the original issues of newspapers and journals which he had received from Mrs. Elly Landauer. Mrs. Landauer, now deceased, provided correspondence and personal items related to the content and meaning of the "Political Papers." We are especially grateful to the Hoover Library, the Harvard-Andover Divinity Library, the Oesterreichische Nationalbibliotek, the Preussischer Kulturbesitz, and the British Library for assistance in locating book reviews and other items that had not been preserved by Landauer. The Homer Babbidge Library of the University of Connecticut and the University of Hartford Library have been continuously helpful. We have requested permission from responsible sources for the use of copyrighted materials. The American Philosophical Society made available the Walter Landauer-L. C. Dunn correspondence, and we appreciate their kindness. We are indebted to Dr. Catherine Skinner, Chair, Publications Committee of the Connecticut Academy of Arts and Sciences for her cooperation, patience and helpful advice in the management of the manuscript

## WALTER LANDAUER
## POLITICAL PAPERS (1919-1929)

The following is a chronological list of all of Landauer's political writings that we were able to trace. The originals and translations are housed with the Landauer papers in the Thomas J. Dodd Research Center at The University of Connecticut. The items selected for publication here are shown in bold type.

| | | | |
|---|---|---|---|
| **20** | **267** | **The Matadors of the "Free German" Youth Movement** | **Die Matadoren der "freideutschen" Jugendbewegung. *Die freie Zeitung* (Bern) IV (11) 2/7/20** |
| **21** | **271** | **A Spiritual Leader of German Youth** | **Ein Geistesführer Jungdeutschlands. *Erkenntnis und Befreiung* (Vienna) II (11) 2/8–14/20** |
| 22 | — | Gustav Landauer's Murderers—Acquitted! | Gustav Landauers Mörder freigesprochen! *Erkenntnis und Befreiung* (Vienna) 2/11-17/20 |
| 23 | — | French teachers as leaders of the Peace Movement | Review of Reinhold Lehmann: Die französischen Volksschullehrer als Schrittmacher der Friedensbewegung. *Erkenntnis und Befreiung (Vienna) II* (39) 8/22-28/20 |
| 24 | — | From the Internationale—Germany | Aus der Internationale—Deutschland. *Erkenntnis und Befreiung* (Vienna) II (41) 9/5-11/20 |
| **25** | **273** | **From Germany—Die Bauernzeitung** | **Aus Deutschland—Die Bauernzeitung. *Der Aufbau* (Zurich) II 45 10/8/20** |
| **26** | **274** | **A mirror to the World** | **Ein Weltspiegel. *Der Aufbau* (Zurich) II 47 10/22/20** |
| 27 | — | Gustav Landauer and the Bavarian Revolution | Gustav Landauer und die bayrische Revolution. *Erkenntnis und Befreiung* (Vienna) II 48 10/24-30/20 |
| **28** | **277** | **Munich impressions** | **Münchener Eindrücke. *Der Aufbau* (Zurich) II (52) 11/26/20** |
| 29 | — | Revolutionizing of the revolutionaries; Intensification of socialism [Review] | Review of Plenge: Revolutionierung... and Vertiefung... *DerAufbau* (Zurich) II (54) 12/10/20 |
| **30** | **280** | **Halle, Kassel, Hanover** | **Halle, Kassel, Hannover. *Der Aufbau* (Zurich) II (55) 12/17/20** |
| **31** | **283** | **On the Problem of Violence** | **Zum Problem der Gewalt. *Der Aufbau* (Zurich) II (56) 12/24/20** |
| 32 | — | Kropotkin's ideas on 'Mutual Aid' and their Meaning | Kropotkins Ideen über die "Gegenseitige Hilfe" und ihre Bedeutung. *Kultur und Fortschritt, No. 558/60* (1920). Felix Dietrich (Gautzsch, bei Leipzig) |
| 33 | — | The "Abend" suppresses the unpalatable truth | Der "Abend" unterdrückt die unliebsame Wahrheit. *Erkenntnis und Befreiung* (Vienna) III (7) 1/9-15/20 |
| 34 | — | Amend: The Old System [Review] | Curt Amend:Das alte System |
| 35 | — | Reimann: The Latrine [Review] and Nordau: Biology of Ethics | Reimann: Die Kloake and Nordau: Biologie der Ethik. *Der Aufbau* (Zurich) II (2) 1/14/21 |
| 36 | — | Beer: British Socialism of Today and Oestreich: Creative Education | M. Beer: Der britische Sozialismus der Gegenwart and Paul Oestreich: Schöpferische Erziehung. *Der Aufbau* (Zurich) II (3) 1/21/21 |
| 37 | — | The German Revolution [Review] | Heinrich Ströbel: Die deutsche Revolution. Ihr Unglück und ihre Rettung. *Der Aufbau* (Zurich) II (4) 1/28/21 |
| 38 | — | Danilevsky: Russia and Europe; and Feitelberg: Paper money Policy in Soviet Russia [Reviews] | N.J. Danilevsky:Russland und Europa; Magnus Feitelberg: Das Papiergeldwesen im Räte-Russland. *Der Aufbau* (Zurich) II (5) 2/4/21 |
| 39 | — | The Papalagi: The speeches of the South Seas chieftain Tuiavii of Tiavea. [Review] | Erich Scheurmann (ed.): Die Reden des Südseehäuptlings Tuiavii aus Tiavea. *Der Aufbau* (Zurich) II 2/11/21 |
| 40 | — | Underground literature in Revolutionary Germany during the World War [Review] | Ernst Drahn & Susanne Leonhard: Unterirdische Literatur im revolutionären Deutschland während des Weltkrieges. *Erkenntnis und Befreiung* (Vienna) III (11) 2/13-19/21 |
| **41** | **285** | **[Obituary of ] Peter Kropotkin** | **Peter Kropotkin. *Der Aufbau* (Zurich) II (9) 3/4/21** |
| 42 | — | Polemics. (Editor introduces letter from W.L. on G. Landauer's views] | Polemisches. *Erkenntnis und Befreiung* (Vienna) III (15) 3/13/21. |

| | | | |
|---|---|---|---|
| 43 | — | On the German Settlements I | Von deutschen Siedlungen I. *Der Aufbau* (Zurich) II (13) 4/1/21 |
| 44 | — | The Revolution in danger! | Die Revolution in Gefahr! *Der Aufbau* (Zurich) II (12) 3/25/21 |
| 45 | — | On the German Settlements II | Von deutschen Siedlungen II. *Der Aufbau* (Zurich) II (13) 4/1/21 |
| 46 | — | On the German Settlements III | Von deutschen Siedlungen III. *Der* (14) 4/1/21 |
| 47 | — | State and Marxism [Review] | Friedrich Lenz: Staat und Marxismus. *Der Aufbau* (Zurich) II (14) 4/8/21 |
| 48 | — | Hegel and the State [Review] | Franz Rosenzweig: Hegel und der Staat. *Der Aufbau* (Zurich) II (14) 4/8/21 |
| 49 | — | On the German Settlements IV | Von deutschen Siedlungen IV. *Der Aufbau* (Zurich) (15) 4/15/21 |
| 50 | — | The Oculi Fables [Review] | Edwin Hoernle: Die Oculi-Fabeln. *Der Aufbau* (Zurich) II (15) 4/15/21 |
| 51 | — | On the German Settlements V | Von deutschen Siedlungen V. *Der Aufbau* (Zurich) II (16) 4/22/21 |
| 52 | — | Jean-Jacques Rousseau [Review] | Henriette Roland-Holst: Jean-Jacques Rousseau. *Der Aufbau* (Zurich) II (17) 4/29/21 |
| 53 | — | On the German Settlements VI (Conclusion) | Von deutschen Siedlungen VI (Schluss). *Der Aufbau* (Zurich) (17) 4/29/21 |
| 54 | — | The Mission of Science [Review] | Erich von Kahler: Der Beruf der Wissenschaft. *Der Aufbau* (Zurich) II (18) 5/13/21 |
| 55 | — | Romantic Socialism [Review] | Siegmund Rubinstein: Romantischer Sozialismus. *Der Aufbau* (Zurich) II (20) 5/5/21 |
| 56 | — | Oestreich: On Trade Schools (Decisive School Reform III); and Kawerau: Sociolological Foundations of Instruction in History | Paul Oestreich (ed.) Zur Produktionsschule (Entschiedene Schulreform III); Siegfried Kawerau: Soziologischer Aufbau des Geschichtsunterrichts. *Der Aufbau* (Zurich) II (21) 5/27/21 |
| 57 | — | How do the worker and the peasant live in Russia and the Ukraine? [Review] | Augustin Souchy: Wie lebt der Arbeiter und Bauer in Russland und in der Ukraine? *Der Aufbau* (Zurich) II (22) 6/3/21 |
| 58 | — | Letters from Exile [Review] | Douglas Goldring: Briefe aus der Verbannung. *Der Aufbau* (Zurich) II (24) 6/17/21 |
| **59** | **287** | **Production Cooperative, Consumer Cooperative, and Settlement** | **Produktivegenossenschaft, Konsumgenossenschaft und Siedlung. *Der Aufbau* (Zurich) II (25) 6/24/21** |
| 60 | — | The Marriage of War and Revolution [Review] | Eugen Rosenstock: Die Hochzeit des Krieges und der Revolution. *Der Aufbau* (Zurich) II (26) 7/1/21 |
| **61** | **291** | **Lenin: Bolshevism and Capitalism** | **Lenin: Bolschewismus und Kapitalismus. *Der Aufbau* (Zurich) II (27) 7/8/21** |
| 62 | — | Ghent behind the lines: Searchlight on the collapse, Vol. I [Review] | Heinrich Wandt: Etappe Gent: Streiflichter zum Zusammenbruch, Band I. *Der Aufbau* (Zurich) II (27) 7/8/21 |
| 63 | — | The stricken Hölderlin [Review] | Erich Trummler (ed.): Der kranke Hölderlin. *Der Aufbau* (Zurich) II (27) 7/8/21 |
| 64 | — | The Brass Check, a study of American Journalism [Review] | Upton Sinclair: Der Sündenlohn. *Der Aufbau* (Zurich) II (28) 7/15/21 |
| 65 | — | Towards new ideals of love [Review] | Olga Mišar: Neuen Liebesidealen entgegen. *Erkenntnis und Befreiung* (Vienna) III (33) 7/17-23/21 |
| 66 | — | Hölderlin: Poems[Review] | Erich Lichtenstein (ed.): Friedrich Hölderlin, Gedichte. *Der Aufbau* (Zurich) II (29) 7/22/21 |
| 67 | — | Gorky:Recollections of L.N. Tolstoy; and Weisengrün: | Maxim Gorky: Erinnerungen an Lev Nikolaievitch Tolstoy; and Paul Weisengrün: Kulturpolitik, Welt- |

| | | | |
|---|---|---|---|
| 68 | — | Cultural policy, world war and socialism. [Reviews] | krieg und Sozialismus. *Erkenntnis und Befreiung* (Vienna) III (35) 7/31 - 8/6/21 |
| 69 | — | The labor cooperative as a form of free socialization. [Review] | E.K. Kloss: Die Arbeitsgenossenschaft als freie Sozializierungsform. *Der Aufbau* (Zurich) II (31) 8/5/21 |
| 70 | — | Hennings: The Stigma; and Grosz: The face of the ruling class. [Reviews] | Emmy Hennings: Das Brandmal; and Georg Grosz: Das Gesicht der herrschenden Klasse. *Erkenntnis und Befreiung* (Vienna) III (36) 8/7-13/21 |
| **71** | **293** | **The Realm of the Antichrist** | **Das Reich des Antichrist. *Der Aufbau* (Zurich) II (32), 8/12/21** |
| 72 | — | Cultural History of Antiquity in Germany and France, and of Germany in the early Middle Ages. [Review] | Rudolf Goette: Kulturgeschichte der Urzeit Germaniens, des Frankreich, und von Deutschland im frühen Mittelalter. *Der Aufbau* (Zurich) II (32) 8/12/21 |
| 73 | — | My stay in Berlin. [Review] | Voltaire: Mein Aufenthalt in Berlin. *Der Aufbau* (Zurich) II (32) 8/12/21 |
| **74** | **296** | **The true face, World History of Socialist Thought. [Review]** | **Alexander von Gleichen-Russwurm: Das wahre Gesicht, Weltgeschichte des sozialistischen Gedankens. *Der Aufbau* (Zurich) II (33) 8/19/21** |
| **75** | **298** | **Anarchy or Authority? [Review]** | **Wilhelm Marr: Anarchie oder Autorität? *Erkenntnis und Befreiung* (Vienna) III (38) 8/28- 9/3/21** |
| 76 | — | On the murder of Erzberger | Zur Ermordung Erzbergers. *Der Aufbau* (Zurich) II (34) 9/2/21 |
| **77** | **299** | **Introduction to the Russian Novel. [Review]** | **Karl Nötzel: Einführung zum russischen Roman. *Der Aufbau* II (34) 9/2/21** |
| 78 | — | von Stein: History of the social movement in France from 1789 to the present, Vol. I: The concept of society and the social history of the French revolution until 1830. Kautsky: Race and Jewry. [Reviews] | Lorenz von Stein: Geschichte der sozialen Bewegung von 1789 bis auf unsere Tage. Band I: Der Begriff der Gesellschaft und die soziale Geschichte der französischen Revolution bis zum Jahre 1830. Karl Kautsky: Rasse und Judentum. *Der Aufbau* (Zurich) II (36) 9/9/21 |
| 79 | — | "Lenin" [Review] | A. Chavash: "Lenin". *Erkenntnis und Befreiung* (Vienna) III (40) 9/18-24/21 |
| 80 | — | Francé: Munich, the laws of a city's life; and The Scale of Life. Wegner: In the house of happiness. Friedrichs: History of Russian Literature. | Raoul H. Francé: München, Die Lebensgesetze einer Stadt; and Die Wage des Lebens. Armin J. Wegner: Im Hause der Glückseligkeit. Ernst Friedrichs: Russische Literaturgeschichte. *Der Aufbau* (Zurich) II (41) 10/14/21 |
| 81 | — | Modern Communistic Communities. [Review] | Michael Tugan-Baranowsky: Die kommunistischen Gemeinwesen der Neuzeit. *Der Aufbau* (Zurich) II (43) 10/25/21 |
| 82 | — | The earth transformed. A book of covenant. [Review] | Max Hayek (ed.): Die verwandelte Erde. Ein Buch der Verheissung. *Erkenntnis und Befreiung* (Vienna) III (46) [n.d., probably 10/30-11/5/21 |
| 83 | — | Sadhana: The path to perfection. [Review] | Rabindranath Tagore: Das Pfad zur Vollendung. *Der Aufbau* (Zurich) II (45) 11/11/21 |
| 84 | — | How a revolution was destroyed. [Review] | Eduard Bernstein: Wie eine Revolution zu Grunde ging. *Erkenntnis und Befreiung* (Vienna) III (48) [n.d., probably 11/13-19/21 |
| 85 | — | Socialist Bookshelf. [Review] | Karl Nötzel: Sozialistische Bücherei. *Der Aufbau* (Zurich) II (51) 12/23/21 |
| **86** | **301** | **Socialist Literature** | **Sozialistische Dichtung. *Der Aufbau* (Zurich) III (8) 2/24/22** |

| | | | |
|---|---|---|---|
| **113** | **345** | **Wehrli, Gerber: Henry Ford again (To Mr. Walter Landauer) (Observations by W. Wehrli and M. Gerber, ed.)** | **Nochmals Henry Ford (An Herrn Walter Landauer) (Bemerkungen von W. Wehrli und M. Gerber, Red.) *Der Aufbau* (Zurich) V (24) 6/13/24** |
| **114** | **347** | **Conspiracy and Murder in Post-War German Life** | **Verschwörung und Mord im deutschen Staatsleben der Nachkriegszeit. *Die Menschheit* XI (22) 8/1/24** |
| **115** | **348** | **The Prosecution of Dr. J. E. Gumbel** | **Das Verfahren gegen Dr. J.E. Gumbel. *Die Menschheit* (Wiesbaden) IX (24) 8/15/24** |
| **116** | **350** | **The Disciplining of Gumbel** | **Die Massregelung Gumbels. *Die Sonntags-Zeitung* (Heilbronn) V (33) 8/17/24** |
| **117** | **351** | **On the Gumbel Case** | **Zum Fall Gumbel. *Volkszeitung* (Heidelberg) VI (191) 8/19/24** |
| **118** | **352** | **Summons to Murder** | **Aufruf zum Mord. *Die Sonntags-Zeitung* (Heilbronn) V (34) 8/24/24** |
| 119 | — | A. Bietenholz-Gerhard: Henry Ford again | Nochmals Henry Ford. *Der Aufbau* (Zurich) V (35) 8/29/24 |
| 120 | — | Henry Ford | Henry Ford. *Die Gemeinwirtschaft* (Greiz) IV (9) 9/9/24, pp. 197-201 |
| 121 | — | Why Henry Ford is voting for Coolidge | Warum Henry Ford für Coolidge wählt. *Der Aufbau* (Zurich) V (52) 12/26/24 |
| 122 | — | The Farmers' Cooperative Movement in the U.S.A. | Über die Kooperativbewegung der Farmer in den Vereinigten Staaten von Amerika. *Die Gemeinwirtschaft* (Greiz) Nr. 3, March 1925, pp.75-78 |
| 123 | — | The same, part II | Derselbe, II. Teil; *Die Gemeinwirtschaft* (Greiz) Nr. 4, April 1925, pp.92-96 |
| **124** | **353** | **Freedom of the Individual in Popper's Guaranteed Subsistence for All** | **Die Freiheit des Individuums in Poppers Nährpflicht programm. *Mitteilungen des Vereins "Allgemeine Nährpflicht"* (Frankfurt a.M.) May 1925** |
| **125** | **358** | **The Case of Gumbel** | **Der Fall Gumbel. *Der Aufbau* (Lucerne) VI (27) 7/3/25** |
| 126 | — | The Lives of Others | Das Leben der Andern. *Der Aufbau* (Lucerne) VI (33) 8/14/25 |
| 127 | — | Dayton, Prohibition, and Ku Klux Klan | Dayton, Prohibition, und Ku Klux Klan. *Sonntags-Zeitung* (Stuttgart) VII (2) 1/10/26 |
| 128 | — | A grand coalition? | Grosse Koalition? *Die Sonntags-Zeitung* (Stuttgart) VII (2) 1/10/26 |
| 129 | — | Arthur Nash | Arthur Nash. *Die Sonntags-Zeitung* (Stuttgart) VII (9) 2/28/26 |
| 130 | — | Consequences of the alcohol policy in North America | Die Folgen der Alkoholpolitik in Nordamerika. *Die Gemeinwirtschaft* (Hermsdorf) VI (3) 3/26/26 |
| 131 | — | Crime as destiny | Verbrechen als Schicksal. *Die Menschheit* (Wiesbaden) XVI (19) 5/5/29 |

# INTRODUCTION

## CHAPTER ONE

### The Landauer Political Papers

The abrupt termination of the war in November, 1918 brought cataclysmic changes to Germany, among them the abdication of Kaiser Wilhelm II, the end of the German Empire, and a future rendered chaotic by uninterrupted social, economic and political crises. The enormity of the problems, separate and aggregate, and the endless obligations and opportunities which challenged the new Weimar Republic demanded a public accounting. A revolution was in progress. We are fortunate to have detailed reports from a young and keen-eyed citizen, a student of science in the University of Heidelberg: the political papers of Walter Landauer. His observations shed light on aspects of the war just concluded and on the reconstruction of a devastated nation. Seventy million people traumatized by unexpected military defeat were now charged with repair of structural damage and design of a democratic government that could meet the challenges of the future.

The written observations are not those of an unbiased reporter—they are fearless and unequivocal opinions of a concerned citizen, a pacifist opposed to violence in any form, with strong convictions about the importance of the individual in society. He was an idealist in a world beset by problems that cried out for immediate and pragmatic solutions. The documents are rare because the journals in which they were published were sometimes suppressed; there are few holdings in either European or American libraries, and none have been translated. Landauer wrote openly about:

- public guilt for "Germany's war"
- the lack of morality in government
- the pusillanimous role of the intelligentsia
- the need for true democracy
- the evils of the party system of government
- the iniquities of capitalism
- the virtues of socialism and the means to achieve it
- a system of guaranteed subsistence for all citizens
- injustice in the courts and the parasitic nature of the military

The Landauer Papers were originally published in the German language, predominantly in socialist journals, now defunct, in Switzerland, Austria and Germany. Because of the insights they provide into German post-World War I politics, economy and social problems, the papers have been translated into English for the convenience of English-speaking students and interested general readers. The papers deal particularly with the Weimar period of German history, but he comments as well on certain affairs of America (e.g., Henry Ford, Prohibition) and on several aspects of Russian literature and politics. We believe that the travails of Germany in the Hitler era invite a close inspection of the Landauer Papers. Indeed, some observations of the Weimar period give indications of activities conducive to the rise of Hitler's influence. In Weimar Germany there was a political fluidity in the sense of party creation and coalitions which played a role in augmenting the political instability of a wide range of German citizens both left and right. Eventually a formidable right-wing bloc could be recognized.

**Walter Landauer in the WWI service of the International Red Cross**

Landauer's observations on the social and economic relationships between people had led him to adopt socialism as a means to achieve optimal social benefits, fundamentally based on equal rights and equal opportunity for all. He had studied its history in Germany and elsewhere, and now, even as a young man, he had decided that the people of Germany should incorporate the principles of socialism. This ambition, however unrealistic, bolstered by antipathy for capitalism, guided much of his social and political thinking.

Landauer was opposed to violence in any form, on any scale. He was disturbed by both the governmental preparation for war and the enthusiastic approval of that policy by the German people. In 1915, at age 19, having completed his Abitur, he was admitted to the University of Heidelberg. He was eligible for service in the armed forces, but as a pacifist and conscientious objector he was given leave from the University and was assigned to the

International Red Cross. He served for two years in hospitals in France and Poland. This service reaffirmed his opposition to violence and to war; violence and killing are unconscionable instruments of injustice, he reasoned.

He saw in a centralized government a device for enslavement of the governed, for unreasonable and fruitless control over the individual's rights and prevention of his ability to attain or even aspire to his potential.

We look at the details of Landauer's early life to find a rationale for his attention to public issues.

## The Landauer Family

The Landauers (see Appendix for the family tree) had settled in Baden; Walter's grandparents, Hermann and Rosa, had established a retail footwear business in Karlsruhe. There were three sons: Felix, who joined in his father's business; Friedrich, who studied law, and Gustav, who was a prolific writer and champion of the cause of socialism. On both counts he influenced his nephew during his formative years. Friedrich, at thirty years of age, had been appointed District Judge in Mannheim, an exceptional position for a man of Jewish ancestry, and one which offered economic security and opportunity, respectability and recognition in the community. He married Charlotte (Lotte) Ziegler in 1894; their first son, Hermann, was born in Phillipsburg prior to the judicial appointment. He was inclined toward a career in business which took him, as a refugee, to Chile in 1938 and to the U.S.A. in 1946.

Walter, second son of Friedrich, was born in Mannheim, July 15, 1896. Friedrich's promising career ended prematurely with his death in May, 1901. Two years later Charlotte married Hugo Landauer, a cousin of Friedrich. Hugo was a merchant who had moved many times and settled finally in Karlsruhe, where he established a watch factory. He was financially successful and found it possible to purchase a second family residence, a fruit and nut farm, on Lake Constance. He was called upon repeatedly, because of his inclination toward socialism, to provide financial assistance to Gustav, Walter's uncle.

The family included Heinz, who was mentally retarded and later was a victim of the Nazi euthanasia program. A year later Erich was born; he was a healthy, cheerful child but, almost 10 years younger than Walter, he was not really a companion as Walter turned his attention to serious issues. Walter's only sister, Liesel, died in an automobile accident at age of 19; Fritz, in due course went to China where he became a victim of drug addiction. At the request of his daughter, he was returned to Germany where he received institutional care until his death in 1960.

Walter had changed residence several times during his elementary school years because of the requirements of his stepfather's business career. He adapted quickly, learned easily, and in due course was adequately prepared for the change of pace when he enrolled in the Reform Gymnasium, Goethe School, Karlsruhe.

He liked the atmosphere, the variety of subject matter, the limitless challenge. He liked language—not only German grammar, structure and ideas, but also Latin, French and English. He was fascinated by the social problems of other countries and with their attempts to solve them. He thought seriously about the forces which created problems for whole societies, and he discussed these questions with his classmates. Not uncommonly, the writings of Gustav Landauer helped to illuminate the complex questions of socialism. As a youth Walter had watched Germany's industrialization and diversion of her industrial might to the tools of war on land and sea, unmistakable preparation for armed conflict.

Walter, now sixteen years of age, had established the intellectual habit of exploring, analyzing, reaching conclusions, taking positions and defending them in debate. He took pride in his independence; he enjoyed the world of scholarship. The gymnasium students were introduced to the rich tradition of German literature and its treasures. Goethe was an endless resource of poetry, drama, and essays in the natural sciences. They became acquainted with the poetry of Schiller, of Hölderlin and Heine, and more contemporary work of Richard Dehmel, a close friend of Hedwig Lachmann, Gustav's wife, and of Hermann Hesse: seeing in poetry a rhythmic and almost musical way of expressing emotion and profound truth. This was a new world with rich rewards. By the time he had completed his studies at the Reform Gymnasium, Goethe School, in Karlsruhe, he had reached some rather fundamental societal conclusions. Paramount in his thinking was the intrinsic, immeasurably great value of the individual, and its corollary, the importance to society of realizing the full potential of all individuals.

Walter, emboldened by his intellectual skirmishes with his peers, thought seriously about a career in law, continuing the abbreviated career of his father. But there were other possibilities. His imagination ranged over the many choices. He was most attracted to biology, an active field full of promise. He reasoned that biological investigations are destined to make significant and lasting contributions to humanity, particularly in the practice of medicine. Finally, he concluded, a professor's life is appealing— independence, opportunity to make important contributions to society, limitless challenge, perennial association with minds equally dedicated to truth. The decision, at first tentative and uncertain, invited serious consideration, and purposeful reading which approached actual preparation. There was a sense of maturity, a certain satisfaction even in the preliminary definition of a professional goal.

Landauer, like all biologists of the day, was impressed by Darwin's concept of organic evolution and the role of natural selection in the process. Certainly humans, like all other living creatures, were bound by the laws of inheritance. But Landauer did not advocate guidance of human evolution by selection of human breeding stock. He was serious, however, about promoting social and political practices which would lead to maximal development of the talents of every individual.

Landauer's matriculation in the University of Heidelberg was delayed until completion of War Service, December, 1918. In 1924 he seized the opportunity to pursue a professional biology career in America. He retired from the University of Connecticut in 1964; grateful for the support which the University had given him for 40 years, he presented to the University his large collection of Kaethe Kollwitz prints. His career in science is reviewed by Clark and Pierro (1994).

## Uncle Gustav

In 1896, the year of Walter Landauer's birth, Gustav was 26 years of age, and he presented to the "London International Congress" a lecture on "Social Democracy in Germany". The lecture, privately published, is the first of 27 items in a collection made by Walter, "Varia by and about G.L.", now in the Landauer Archives, Dodd Research Center, University of Connecticut. The fact that the collection was made suggests the respect, the love, perhaps even the reverence, of nephew for uncle. The lecture recognizes "the political maturity and free temper" of England and France and it castigates Germany "where huge masses are wont to dance to the tune played by the upper regions of the party government. To understand this we must remember that Germany enjoys the doubtful honor of being the home of monarchism and militarism".

Gustav Landauer (1870-1919)

Gustav became a leader for the cause of socialism. For a time, beginning in 1893, he had edited Der Sozialist. He was forthright and outspoken, with ideas sufficiently at odds with the mainstream of thought that he was arrested and imprisoned in 1893 and many times thereafter. In 1908 he wrote *The Twelve Articles of the Socialist Federation* which declared the intention of creating "independently managed economic communities trading together in justice" which would "replace governments and the capitalist economy". He expanded these Articles in a lecture which was published three years later under the title, *For Socialism. (Aufruf zum Sozialismus, Second Edition. Paul Cassirer Verlag, Berlin. 1919.)*. It was a forceful call to action for Federation members and for those who could be persuaded to join. The preface to the second edition begins, "The revolution has come...", an exaggeration which meant only that, with the end of the war, the opportunity to push forward was at hand. One of his biographers, Eugene Lunn, calls Gustav "Prophet of Community". He also wrote extensively about literature and the arts, and he lectured to appreciative audiences in Berlin and elsewhere on Shakespeare, Walt Whitman, and Oscar Wilde.

Walter was moved by Gustav's enthusiasm and persistence and by the range of his interests. On February 10, 1918, Walter met with his uncle and Hedwig Lachmann, Gustav's second wife. There is no record of their discussion, but certainly they reviewed the looming crisis in military affairs. With the end of the war there would not only be need for major change, but opportunities to steer the process. However, neither dared to predict the end of the monarchy; socialistic changes would need to be opportunistic, systematic, and subtle.

Walter recognized that there was much to do, and the feeling of impotence as a citizen strengthened his resolve to oppose centralized political power. Gustav's repeated advocacy of community suggested a pragmatic approach to a political ideal. The visit with Gustav may have given Walter the confidence to support major social and political change which set the tone of his political essays. His observation of political affairs was not so much a civic responsibility as a psychological compulsion.

## Landauer looks at Germany

In the following pages we report Landauer's views on issues of major moment in Germany at the end of World War I: War Guilt, Revolution, Justice.

The issues were not addressed in a systematic, predetermined order, but rather as current events seemed to dictate. There is no evidence that Landauer considered the comment on a particular topic a final pronouncement. Pride of authorship is patent in his collection and preservation of all the original publications and certain related material. His comments on an immediate target usually reflected opinions which had been gestating over a period of time. They tumble forth spontaneously as new situations arise, commonly inspired and amplified by current reading. He gratefully acknowledges such sources, which serve to describe the maturation and changing scope of his opinions.

Landauer published his first essay in *Die freie Zeitung*, January 1919. He criticizes the posture of Heidelberg professors, cites the crimes of the German monarchy, and further explains that "the former enemies of Germany—to whom in our inmost thoughts we should be grateful for the liberation they have brought us—should have doubts about the honor of the German people." Several other pieces were subsequently published in *Die freie Zeitung*, for example in a eulogy of Kurt Eisner whom he compares to President Wilson, he claims that "freedom of thought, of speech and of action have definitely disappeared from Germany's precincts." Similar broadsides against other aspects of German practice and policy are itemized in papers 3, 5, 8 and 12. The last of these appeared in February, 1920. In March, 1920, Hermann Popert devoted an issue of *Der Vortrupp* to "The Masqueraders", a fully documented analysis of the editorial policy and *raison d'être* of *Die freie Zeitung*: *"A press organ for advancing the interests of the Entente, in particular those of France, and damaging the interests of Germany."* Popert lists 16 German contributors during the five years which he studied, among whom is Walter Landauer. Popert reports that "during the war" he repeatedly submitted to the German authorities the evidence of "smuggling of the paper across the Swiss border." He concludes that "the Swiss government, which has always been uniformly loyal to our Fatherland, ... will eventually refuse to tolerate an enterprise and guests who are professional practitioners of high treason against the German State, neighbor and friend of Switzerland." We inquired about the fate of *Die freie Zeitung* of the Bern Landes Bibliothek, which reported that it was not published in Switzerland after March 1920, but was presumably published in another country.

## The Weimar Republic: Bridge to Nazism

Capitulation of Germany's armed forces in the fall of 1918 led to revolutionary changes within Germany and in Germany's relationship with other countries. Because we believe that conditions in postwar Germany invited political proselytizing of the population, we describe them in some detail as a guide to understanding the political stalemate which arose between government representatives of the left and the right on the issues of socialism or return to the monarchy. Intragovernmental strife virtually immobilized reconstruction and rational solutions of other necessary activities. The Landauer Papers provide evidence of the susceptibility of a bewildered nation to the propaganda of the National Socialist German Workers Party (NSDAP).

In the 20th century a united Germany had achieved eminence in science, art and literature; its educational system was a model among progressive western nations. There was much to be proud of. Germany entered World War I with confidence and assurance. The decision of the government was endorsed with the strong and noisy support of Germany's people. It would be a short war, a jubilant victory. After four years of sacrifice and pain and bitter fighting, the people were informed that Germany's armed forces could not continue. The illusion of imminent victory was turned to the shock of defeat for the millions who had worked so hard and so hopefully. The shock was only the beginning of Germany's postwar troubles.

In addition to the uncontrollable consequences of the termination of the monarchy, other events occurred. For example, sailors mutinied in the North Sea ports, and this action by a segment of the military triggered a wave of independent, uncoordinated clusters of action and autonomy throughout Germany in the form of Workers and Soldiers Councils, each claiming administrative responsibilities. The "Councils" were in part an adaptation of the Russian "Soviets" to the conditions and inclinations of the returning German military and anxious native sons. The feverish political activities in the national government and state hierarchies, and locally in the Councils, symbolized the urgency of social and economic problems. Coordination was virtually impossible.

In 1919 a lingering nostalgia for the monarchy persisted. This was not an organized revolt against the republic, but a symptom of uneasiness and anxiety manifested later (March, 1920) in the aborted Kapp putsch. The possibility of a thorough overhauling, like Russia's 1917 revolution, was kept alive by the German Communist Party and sympathetic German leftists. Landauer kept a close eye on events, hoping for changes that would satisfy his democratic ideals. His writings reveal a sense of concern unique among his student contemporaries. He recognized the nationalistic and even racialist efforts of the right. He countered with bold criticism of rule by force, commonly tainted by rank injustice.

The end of the war brought into focus the urgency of rebuilding a nation shattered in substance and in spirit. A great wave of anger was turned against the ruling oligarchy. The atmosphere was confusing and volatile. The magnitude of the task was frustrating. The military *regime* was ended. Germany's resources, for four years dedicated to war, had now to be converted to peacetime economy. The trauma of defeat was accompanied by a sense of futility. Shelter for the masses and for returning soldiers could not be postponed. Jobs that had been necessary, indeed ordained, for the military had to be converted to civilian service. There was massive unemployment. Finance for local, state and federal reconstruction was not available. The Versailles Treaty included burdensome reparation payments dictated by the Allies. German anger, superimposed on inadequate organization, was aimed even at reconstruction efforts.

The uncertainties of the early days of the republic resulted in two kinds of shifting allegiances: political, i.e., parliamentary control, and private clubs and associations for various purposes. By the end of the decade there were almost 40 political parties; the number of nonpolitical groups was much larger. Landauer was aware of the private clubs and their activities and of the infighting in the halls of government for an edge in the continuing struggle for power. The people were confused, frustrated, often angry, and worse, cynical and pessimistic.

Without a solid base for public finance, monetary inflation grew to meaningless proportions; Pinson (1954) comments, "The mad days of inflation brought economic ruin, a loosening of morals and social ties, a wave of political and social malcontent and cynicism, and a class of proletarianized burghers who became the happy recruiting ground for all the nationalistic and racialist movements that came to undermine the republic."

The republic, struggling against continuous reminders of the monarchy, pejorative treatment by the victorious Allies, uncontrollable economic crises, communist agitation and even defects in the constitution, did well to survive. The NSDAP, like other political parties, competed for recognition, without notable success. It sought leverage through a volkish approach.

Among the factors which scholars believe to be important in interpreting German history, one has been repeatedly cited: people's pride in their own history. The future is tempered by both genetic and cultural heritage, the role of "volkism" in determining behavior (Stern, 1961; Mosse, 1964; Butler, 1942). Awareness of an illustrious past applied equally to the millions of men and women who stayed at home and the hundreds of thousands who were now returning from the front lines. All were now united in wanting a Germany that would provide relief from personal deprivation and an active role in the rebirth of a once great nation. Landauer (#10) acknowledges the "slow and steady evolution of human history", but he sees "alongside that, discontinuous leaps in the advance of civilization" through cultural changes.

## Nationalism

The suffering, the destruction and disorder wrought by the war had the effect of focusing on a return to normalcy, a Germany of national goals, national distinction in sports, in science and in the arts. The concept of nationalism revived thoughts of racial strength and purity. The accomplishments of Jews increased their visibility. They were regarded by many as a nonteutonic race, not simply as practitioners of a separate religion. Antisemitism was so frequently observed that it extended beyond personal prejudice. It was a political reality. (Walter Landauer had earned his doctorate in genetics at the University of Heidelberg in 1922. He was nominated for a position in America by one of the world's outstanding geneticists, Richard B. Goldschmidt, then at the Kaiser Wilhelm Institute in Berlin. Landauer accepted the position in 1924 and said later that "there was not room in Germany for two Jewish geneticists.")

## Bavaria

At this time, in 1919, a disturbance was unfolding in the South, centered in Munich. During a visit to Munich in 1920 (#28) he was struck by the often heard greeting "Heil, Aryan". Some of Munich's problems were initiated and perpetuated by Kurt Eisner, a native of Berlin, whose journalistic career brought him to Munich and involved him in Bavarian politics. His activities led him to promote a general strike, which in 1917 was treasonable. He was imprisoned in January (1918) and, when released October 14, returned to Munich. He worked hard, particularly with the Soldiers and Workmen's Councils. On November 7 he became the "provisional" Prime Minister of the Bavarian Republic. It was a stormy tenure, and on February 21, as he prepared to resign, he was assassinated by Count Anton Valley Arco (Mitchell, 1965). Landauer (#3) eulogized: "Kurt Eisner wanted to be to the German people what Wilson wanted to be to all nations."

**Eugen Leviné (1883-1919)**

Eisner's death triggered weeks of discord in the Bavarian government. Arguments created no consensus. The choice lay essentially between administration by Central Council or by Soviet Republic. After an all-night session on April 4, the assembled spokesmen voted to create the Soviet Republic of Bavaria. It came into being on April 7 and lasted for six days under the leadership of Eugen Leviné. Gustav Landauer had been induced to join Eisner in the Bavarian adventure, and he participated ably. (Walter had marked Gustav's speeches in the Verhandlungen des provisorischen Nationalrates des Volksstaates Bayern im Jahre 1918-1919 [Proceedings of the Provisional National Council of the

People's State of Bavaria in 1918-1919], now in the Landauer Archives, Dodd Center, University of Connecticut.) His involvement in the Bavarian Revolution was too prominent to be ignored. Mitchell (1965) reports that Gustav, in a postcard to Fritz Mauthner, said "I am now commissar for propaganda, education, science, art and a few other things." After two weeks of confusion, Johannes Hoffmann, the previous Prime Minister, called upon the federal government for armed support. On May 2 a contingent of Freikorps assaulted and murdered a group of already imprisoned leaders of the Soviet Republic. Again Walter Landauer was touched by tragedy. His uncle, Gustav, was among the murdered.

## Hitler and Nazism

Hitler was a native of Austria, born of middle-class parents. His father ruled his house- hold with a firm hand, perhaps literally, and Adolf's childhood may best be described as anxious. After the death of his father (1903) and his mother (1907) Adolf, now 18 years of age, went to Vienna to study architecture. He had not distinguished himself in either elementary school or in gynmasium. His only recognizable talent was art, and indeed, for a time he earned a sort of hand-to-mouth living, painting tourist attractions in Vienna. He was not accepted in architecture school. In 1913, he went to Munich and enrolled in a course in military training; he was accepted by the army and served honorably for four years as a messenger, was wounded twice and was awarded the Iron Cross. When he was released from the hospital in 1919 he returned to Munich, still officially in military service. Kershaw (1998) reports that at that time there were 73 *voelkisch* groups in Germany (about 15 in Munich) with a combined membership of 200,000. This is the climate which prevailed as Hitler returned from military duty. On September 12, he was directed to attend a meeting of the German Workers Party. It proved to be a momentous event. He entered the debate with such understanding and intensity that he impressed the chairman of the meeting, Anton Drexler, who discussed membership with him. Hitler joined the committee and at once became active as a speaker, and soon as Recruitment Director. In 1920 the German Workers Party was reconstituted as the National Socialist German Workers Party (NSDAP). Hitler assumed a position of prominence in the Party, attracted financial support from Munich's wealthy leaders, bought a newspaper, the *Voelkischer Beobachter*, and became well known in other parts of Germany. In July, 1921, he resigned from the party and returned only on the condition that he be given complete control. This evidence of Hitler's persuasive manner led to further meetings and larger audiences. His success was noted now beyond Munich, and he became an "indispensable drummer" for the cause of National Socialism. Party leaders conceived a putsch which would take over the central government. It was not well planned, it was bloody, and it was unsuccessful. Hitler was the obvious leader, and there was no dispute about his guilt. The judge, a nationalist supporter, imposed on Hitler a light prison sentence. When Hitler was released a year later his position in the Party was stronger than ever. He now perceived himself not as a "speaker", not as a recruiter, but as a "Fuehrer".

Coincident with Hitler's increase in popularity, President Ebert died. The search for a successor among government regulars was inconclusive. No candidate emerged with a clear majority. Field Marshal Paul von Hindenburg, Germany's military leader in the war, was finally elected President. He was then, in 1925, 78 years of age. For eight years, he managed Germany's affairs capably with the help of Chancellor Bruening. During this period Adolf Hitler had assumed undisputed leadership of the Nazi Party.

Although party growth was indifferent for several years after 1924, in 1930 the Party polled 6.5 million votes (107 Reichstag seats) and in 1932, 14 million votes (230 seats). Clearly in a commanding position the Nazi house organ, *Der Angriff*, proclaimed: "We enter Parliament in order to supply ourselves, in the arsenal of democracy, with its own weapons, to paralyze the Weimar government with its own assistance. If democracy is so stupid as to give us free tickets and salaries for this purpose, that is its own affair. As the wolf bursts into the flock, so we come." (quoted by Butler, 1942)

Six months later, January 30, 1933, Adolf Hitler became Chancellor of the Third German Empire.

## Morality

Landauer watched the ascendancy of National Socialism from America. His concern for the individual, his moral precepts, his opposition to violence are a striking counterpoint to the eventual ruthless and brutal tactics of Hitler. His philosophy is epitomized in his review (#74) of Gleichen-Russwurm's *World History of Socialist Thought*. "It is a fatal stroke when politics and morality are separated from one another." He reiterates this principle many times, e.g. (#5), where he blames the current difficulties of the German people, June 1919, on "a total collapse of all morality."

Interpersonal relations within human societies have been recognized in the realm of ethics from the beginning. Scanlon (1998) discusses the concept of right and wrong at length. He says (p. 245): "The contrast I am drawing here is not between moral philosophy on the one hand and political philosophy on the other, but between very general claims about morality, or about what we owe to each other, and claims about what is right and wrong in specific circumstances (claims about the justice of social institutions being one example)." Similarly, Fromkin (1999) notes the contemporaneous "development of conscience in religion and philosophy" by geographically separated peoples. This moral awakening, he believes, occurred only recently—about 500 years before Christ.

Landauer writes: "the end of past epochs and the beginning of new ones in the history of man are not so apparent to the observer as our historians like to represent ... Collapse and a new beginning are much interwoven. Hardly perceptible by its contemporaries, the new rises up. Perhaps this is happening today, but we are unable to recognize the delicate seed that some day will blossom into our hopes" (#101). "The revolution has all but expired because of the spiritual lethargy of the German people, and we do not know when the revolutionary spirit will flare up anew and bring about great events, a completely altered mode of thought; and it is for this reason that we feel called upon to prepare the way to herald the time that will realize the fulfillment of our ideals" (#10).

Landauer's defense of his colleague, E. J. Gumbel, affords a summary of his principles of morality (#115, #116, #117): the primacy of the individual and the responsibility of a social institution to its component individuals. Landauer addressed the right of free speech, and he publicly denounced the University of Heidelberg for use of its power to dismiss a faculty member for expressing his opinion. (See below.)

For Landauer, capitalism is an omnipresent moral issue. We cite a few instances in addition to the reference in his letter to Herbst (see below). The capitalist economy has bred its own downfall. Highly intense competition has led to an insufferable mistrust among people, and indeed, among nations (#25). He is not deceived by the Russian concept of the cooperative. It advances small capitalism into the foreground. The freedom and rights of cooperatives signify freedom and rights of capitalism (#61). His opposition to the proposed State Inheritance Law (#87) is based in part on his perception that it would promote capitalism and acquisition of wealth. He saw as well in the success of Henry Ford's innovative methods a dehumanizing of the assembly line worker, a high social price to pay for Ford's capitalistic empire. Landauer's initial comments (#110) evoked a lively exchange with *Aufbau* readers, mostly opposing Landauer's view (#111, #112, #113).

# CHAPTER TWO

## War Guilt

Germany had been faced with the question of guilt from the time of the initiation of war in 1914. The government was therefore active in disseminating its own propaganda. This effort perhaps stimulated efforts of private citizens, such as that reported in the first essay (#1).

With the adoption of the constitution and the election of government officials in 1919, a segment of the Department of State included the Commission on War Guilt (*Kriegsschuldreferat*), an organization dedicated to anti-guilt propaganda.

Geiss expresses the opinion that Bernhard von Bülow was sent to Versailles with the express purpose, by opposition to Article 231, of scuttling the entire Peace Treaty. When he returned, unsuccessful in that mission, he was appointed Secretary of State. It was he, therefore, who directed the substantial efforts of the *Kriegsschuldreferat*, which included two main organizations, the *Arbeitsauschuss deutscher Verbaende* [Executive Subcommittee of German Associations] and the *Zentralstelle zur Erforschung der Kriegsschuldfrage* [Central Office to Investigate the Question of War Guilt]. A periodical, *Berliner Monatshefte*, was funded through the *Kriegsschuldreferat*, which also subsidized publication through other outlets. A "host of writers" were commissioned to provide a continuous flow of exculpatory propaganda. Although Landauer could not have been spared the government's propaganda, he did not acknowledge it directly, and he persisted in his accusations of national guilt.

Landauer uses the yardstick of war guilt even for the measurement of individuals. For example, Gustav Wyneken, a leader in the German youth movement, advises, "What a consecration the war is for you!" Landauer comments (# 20) "such utterances do not lag in any respect behind the Manifesto of the 93, or behind Pastor Traub".

Landauer had only turned 18 when war was declared. But he was fully aware through current periodicals and through conversations at school and at home of significant antecedents of the war. His Germany was a collage of outstanding achievements in science, art, drama, literature, and of commendable legislation in social welfare: pensions, unemployment insurance, health insurance. The Prussian premise of class distinction contradicted the ideal of social equality. But Landauer saw, too, industrial growth and the diversion of human and physical resources to preparation for war. These internal changes were inseparably connected to Germany's acquisition of foreign colonies, a policy not wholly unrelated to aspirations for naval forces which might rival those of England. Landauer found much that was satisfying, but the triad of expansionism, nationalism and militarism was loathesome. It conflicted with his deep-seated hatred of violence. There could be no justification for war. He was no less disturbed by the general approval of the people which reached a climax of patriotic euphoria with the declaration of war, and he felt keenly the disregard for the concept of humanity. He was an avowed and deeply committed pacifist, and he felt the sting of public disapproval, "at a time when pacifists were despised in Germany and virtually regarded as traitors to their country" (# 2).

Walter's pacifism and antipathy to violence, and his faith in socialism may have arisen independently from his own experiences; but the congruity with Gustav's social and political positions is striking. Controversy and reassessment of values characterized both. Though Walter respected Gustav and found much to emulate, direct derivation from Gustav's positions is unlikely..

Within weeks of the westward march through Belgium, he saw the Manifesto of the Committee of 93. The Committee included professors of a wide range of specialties from sixteen German universities. "We hereby protest, to the civilized world, against the lies and calumnies with which our enemies are endeavoring to stain the honor of Germany in her hard struggle for existence—in a struggle which has been forced upon her." Disclaimers follow concerning Germany's guilt for having caused the war, trespass in neutral Belgium, devastation of Louvain, disdain for international law. It closes with a strong endorsement of "German militarism" without which "German civilization would long since have been extirpated" (Lutz, 1932).

Although its claims were indefensible, this burst of patriotism is seen as a reasonable caricature of the intellectual elite. Landauer's first political paper singled out 76 Heidelberg professors and docents because they had requested that the People's Government of Baden "use its good offices with the Federal Government to bring to a halt the random circulation of one-sided documents which 'allegedly' admit German guilt for initiating the war." "These reactionary scholars, who should after all be our leaders, are instead dragging the German people into further misery..."

He and his fellow pacifists suffered during the war and even before, "as criminals against the German Nation." And now "they are the ones who find it a paradox or pretend to, if any one speaks at all of German guilt." "What is necessary today once again is to establish what the issue is when we renew over and over what is at stake when we speak of German guilt." (#8)

During the war "we fought against militarism and the war psychosis," "against expansionism and Pan-Germanism." What was needed then and is still needed is ***truth***, he asserts.

From the tumultuous days leading to military armistice, it was hoped that the German people would have disavowed "its criminals and war mongers," that Germany's responsibility for the war would be seen as incontrovertibly self-evident. Landauer, convinced that Germany had been responsible for a cruel, perhaps even criminal war, vainly hoped that the Allies, whom he recognized as saviors of Germany, would be properly rewarded by the Peace Treaty. He found no support for his view either among the Allies or in Germany, but he presents a thoughtful analysis of the situation (#8). He is relentless, however, and blames for the failure of favorable terms not the German negotiators, but "the entire German people."

Until the German people, from within rather than from outside, recognize the spiritual foundation of a new age, what is needed, and an important objective of the revolution, is the "complete transformation of the German mentality." Basic to this change is the acknowledgement of German war guilt.

## The Academic Establishment

Landauer, of course, had personal reasons to look closely at the role of professors in German culture and politics, e.g., his own connection with the University of Heidelberg. But there were institutional reasons as well. Fifty years before, the position of German universities had been unambiguously stated by the rector of the University of Berlin, Emil Du Bois-Reymond in an address to the University in 1870, "The University billeted across from the royal palace is the intellectual bodyguard of the House of Hohenzollern" (Emil Du Bois-Reymond, 1912) Although the House of Hohenzollern now had only historical relevance, the relationship between the government and Germany's universities persisted, intimate and interdependent. The government provided financial support and retained an element of control extending even to approval of professorial appointments. The faculty, however independent in their scholarly duties, were conservative with due respect for governmental policy. Professors were civil servants, and they were expected to support the political *régime*.

Landauer found it convenient to focus on the intellectual elite in his early political observations: partly because of the Committee of 93, partly because of the traditionally conservative university posture, virtually dictated by the administrative ties to the government, and partly because of the professorial imprint on university graduates, particularly in the learned professions. It is significant that Landauer offers Kurt Eisner as a professorial counterpoint.

Landauer persists in his criticism of professors and other intellectuals, notably for their failure to exercise leadership in social reform, but also for their inclination toward nationalism. His comments are serious and on the mark, critical, sometimes accusatory, and sometimes offered with a touch of irony. He writes, for example,(#4) about Councils of Intellectual Workers. During the previous November, Workers' and Soldiers' Councils were created throughout Germany, commonly with governmental executive powers. Councils of Intellectual Workers also were created in many cities. Some, presumably those in the south with their own newspaper, edited by Richard Staudinger in Munich, proposed to go a step farther and form a Party of Intellectual Workers. Landauer pokes fun at the editor, at the content of the proposed paper, and finally impales the membership on the barb:"The question whether more intellect is to be found among officials or among workers is to be posed but not to be answered...." He further cautions: "What one must keep in mind, especially in this time of revolution, is that it was the German intellectuals who were the most effective advocates for the German War, and that even now the majority of them stand as barriers to the development of a more liberated humanity." "It cannot be the task of the intellectuals to breed a new spirit of provincialism in order to impress the population from the top down—that approach is now definitely over! Whoever wants to achieve great things can do so only as part of the people, with the people, in order to realize the watchword: 'We want to be a single people of brothers.'"

It is interesting that Landauer, on the day before acceptance of the Constitution of the Weimar Republic, emphasizes (#9) the revolutionary mission of the *proletariat*, "at whose side only a few members of the 'educated' classes stand in this grave battle." But in this essay Landauer focuses on the failure of professors to assume a position of leadership in the political battles for the rights of individuals. "...We must strike out in wholly new directions." It will be necessary to gather people around us "who are gladly prepared and determined to stand with us in the endlessly difficult task of reconstructing the history of humanity.." This is a tall order which Landauer addresses by categorizing professors according to recognition and dedication to the cause. "First among them are those few who, throughout all the tempests of Pan-German power politics and systematic lying and vandalism, have preserved, despite the most oppressive personal sacrifices, a love of unconditional unqualified truth, and the courage to oppose crime and violence. Professor Nicolai, Professor Foerster and a few, all too few others, are those I have in mind here." Nicolai was a physician and professor of physiology at the University of Berlin. A second, large, group is comprised of those who "honestly tried to remain neutral."

"But we must join open battle with those we call Pan-Germans and Ultra-Pan-Germans, with those who even today find it necessary to rape mankind with the barbarity of their political views." This group is best represented by Houston Stewart Chamberlain who had been particularly active two decades earlier, a strong advocate of the exclusion of Jews to achieve purity of the

German race. And Landauer includes Professor Lenard, Nobel Laureate and professor of Physics, of Heidelberg. He cites others and asks Professor Hermann Kraemer of Hohenheim, "Have you never found even a glimmer of understanding of the profound truth of the words of Theophrastus Paracelsus of Hohenheim: "The arts are not sealed up in some Fatherland, but are distributed throughout the whole world; they are not in some individual or in one place, but must be gathered together and sought out wherever they are." Landauer despairs of bringing these gentlemen to their senses and concludes: "One of the first tasks of Germany, one of the most important duties of our coming (note: July 30, 1919) revolution, must be to find pure educators who are imbued with science and art, and for that very reason religion also, and therefore also deepest love of mankind."

Landauer refers above to the "Pan-Germans and Ultra-Pan-Germans" disparagingly. But he was a member of the Monist League and had read its periodicals faithfully (#2). The Monist League was founded in 1906 after a two-year incubation period following its proposal by Professor Ernst Haeckel (1834-1919), a biologist at the University of Jena. Haeckel was working on his dissertation when Darwin's Origin of Species appeared in 1859. He became one of the great biologists of the world, dedicated and enormously productive in a wide range of zoological disciplines. It is of special interest, however, that he was intrigued by Darwin's theory of natural selection and began at once to adapt the concept to the evolution of man, to the origin of the races of man and, extrapolating, to the future of the German people through the concept of nationalism and racial purity. His passion for a new, strong, expansionist Germany also led him to join in the formation of the Pan-German League whose ambitions included militarism, colonial acquisitions and world domination.

Gasman (1971) says of the Monist League: "The League conceived its work to be the development and fostering of a complete social, cultural, religious and political program for Germany." An official program of the League stated that it "strives to bring together all individuals and organizations who no longer rely on the church for their outlook on the world."

# CHAPTER THREE

## Revolution

Landauer refers many times to the "revolution," often with disappointment and despair, perhaps because his expectations were not clearly formulated. Nevertheless, Germany's revolution of 1918-1919 did encompass changes of great magnitude and long-range significance. Social, political and economic problems demanded immediate solution. It was a perplexing and turbulent time.

For two years, 1916-1918, Landauer's life had been circumscribed by his Red Cross hospital duties. Certainly, he was interested in the progress of the war, above all in its termination. However, there is no reason to believe that he had information beyond what was provided by a rigidly controlled press. His conversation with Gustav in the previous February could not have anticipated the events of autumn 1918.

Now in January 1919, his personal affairs were focused on his professional education. He was anxious for change, but he had no organized political ties to specify or achieve his objectives. No matter how fanciful his hopes, he was limited by training and by inclination to observation, to analysis, to complaint. His ideal of a genuinely socialistic government, with direct participation of all interested citizens, could not be realized at the hands of the governmental leaders even if they called themselves socialists. The entire system of political administration was based on representational government. This principle and the machinery which maintains it, Landauer believed, are predicated on the suppression of individual rights. He sought a substitute. We must presume, therefore, that he had in mind replacement of the central government by a system based on socialistic principles and mutual aid communities.

Regardless of impediments to installation of a socialist *regime*, Landauer watched events closely. "The revolution has failed to achieve a single one of the demands that we must await and insist on today..." Landauer (#5, June 21, 1919) is worried: "Whither Germany is now the fearful question which burdens the souls of all feeling individuals." Those beyond the borders of Germany "will never be able to share the oppressive sense of the total collapse of all morality, of a moral uncertainty bordering on characterlessness of the entire German people as it weighs each of us down, threatening to rob us of the last spark of hope."

"The mentality of the revolutionary leaders, the governing socialists, is too well known to require comment." Landauer appears to acknowledge that the Weimar Republic is the revolution, and, although he indicts the governmental leaders, he blames the German people as well. "We see again with frightful clarity today that throughout the years of national massacres, throughout all the humiliation, lies and violence that the German people found acceptable, nothing has been altered in the thinking of the people or in their attitude toward the principles of justice and morality. The spirit of 1914 has begun to triumph over the revolution!" "All the political parties....pursue one and the same method of joining battle: by the denial of any sort of morality and by a disposition toward the most extreme materialistic basic principles, the mean-spiritedness of their conception of human responsibility, and to some degree perhaps an unconscious denial of all forms of Christianity."

Landauer finds the elected representatives woefully wanting in their guiding principles and their respect for humanity but, he maintains, the basic cause of their deficiencies is the people who elected them. Murder and hatred had been drummed into the German people who were ultimately rescued by "a coalition of nations that was fortunately more powerful." With the military defeat "the revolution was christened." Landauer feels that nothing, either constructive or destructive, has emerged thus far from the revolution. "So one should not be surprised that bestial crimes are committed before our very eyes with self-confident assurance." "In Munich, (Eugen) Leviné condemned to death, and the sentence immediately executed!....an absolutely illegal court murdered a man without right of appeal despite the virtually unanimous favorable testimony of witnesses." And Ernst Toller, "a man so thoroughly pure of character, animated only by the highest ideals" is next. "I charge everyone, and I hope that my appeal to human feelings will not echo ineffectually, to contribute his share to the elimination of militaristic class justice!" In a separate piece (#6, June 21, 1919) Landauer attacks the system bluntly.

He chooses to advocate a revolution which would certainly list among its social objectives many which had been assured in the adoption of the Weimar constitution. With one striking exception—a centralized system for administration of social change. His opposition to a centralized state government was unshakably rooted in his vision of the rights and duties of the individual. Landauer continues to be disturbed:(#7, June 25, 1919) "The revolutionary movement in Germany is approaching the end of a period of development that has not taken any forward step, that has brought us only misery and ruin.... With unchallengeable certainty, with incredibly insolent behavior, the counter-revolution is rising in Germany and is arming itself for a powerful assault." "One cannot emphatically enough stress how large is the share of responsibility for the systematic suppression of a revolution of the spirit that must be borne expressly by socialist parties, capable only of worshipping the slogans of their party idols. The war required "frightful sacrifices" from which we must have learned at least "the curse of force and violence, the curse of hatred and injustice." "Fight against injustice wherever you encounter it, fight for the right and for justice, so that these may be achieved!" "But, brothers, the sacred duty falls to us to assure that barbarism disappears in our own house." "From the military whose essence it is to be parasites on humanity, we cannot expect humane feeling or a sense of justice. The guilt for all these appalling deeds lies with those who still support the military, with those who believe they are necessary for their own security, and with those to whom the actions of the military are welcome."

The military, he argues, sees "in our vanguard their worst enemies," "but there are people who should be called upon to support right and freedom who instead support the military lust for murder, that we cannot and will not tolerate." "What we demand, however, of the Bavarian government is that it immediately take steps to eliminate the military courts..." And, "we call upon all men who are striving for a new era, for a loftier humanity to support our demands in every way."

Landauer appears to have set for himself a difficult goal. He regards the Weimar Republic as a noble objective, replete with many of his own social values, amply worthy of support. But he detects flaws in the actual performance of the government, flaws in the interpretation of the directives embodied in the constitution. He seeks to correct the system by shifting to the people the blame for moral deficiencies of the administration, of the Reichstag, of the courts. That done, he seeks to persuade the people to recognize their guilt, to see the national problems in broad view so that they may be solved. Landauer was not a politician, he was an idealist. Among the premises for his vision of the social order are first, eschewal of force and violence, hence elimination of the military forces in all their ramifications, including judicial; second, abandonment of the party system as a device for selecting the people's representatives in government; third, replacement of centralized political administration with a community-based system of interaction and interdependence of individuals of varied talents dedicated to the achievement of objectives in the common interest; fourth, elimination of a capitalistic economy; fifth, adoption of those civic principles which will enhance the opportunities for individuals to develop their skills and talents in any way that will contribute to the welfare of the community, hence the nation. It is clear that the programs which these premises imply could not be satisfied even by drastic modification of the Weimar constitutional proposal. Landauer, without either political organization or the ambition to create one, had to be satisfied with criticism of the existing structure, timely, perceptive, biting, accurate, but withal, ineffectual.

Landauer freely incorporated the tenets of socialism into his own thinking. He followed the career and principles of Peter Kropotkin with admiration; in a 1921 obituary he recalls a critical, reorienting decision: "Either one acts falsely against oneself and says let mankind perish so long as I can flourish! Or one becomes a socialist and aims at the restructuring of society" (#41).

Landauer pondered the schism between political reality and his own concept of a functioning socialistic system. His thoughts were crystallized by two unrelated items, one promoting a socialized economy, and the other a socialist program.

The first is a draft bill on "State Inheritance Law and the Cooperative Movement" (#87) which would provide that "the State is the universal legateee of its citizens," and concurrently a State Estate Bank with power to transfer estate titles. The net result of the two laws would be "gradually an automatic socialization of economic life." Landauer notes that "the harsh lessons of recent years (especially in Russia) and the bitter distress that still faces us, will compel more and more people, even in Marxist parties, to recognize what individual socialists outside the parties (and therefore without influence) have long seen: that all attempts to socialize the economy will sooner or later fail until people become socialist in their utmost beliefs, thoughts and actions; until their socialism is effective in action and life (within their families and sphere of influence) rather than expressed in theory and propaganda. That this is not the case today hardly anyone would dare to deny. We shall therefore have to return to the point that the socialization of people must be placed in the foreground of our efforts." "It should be clear from the foregoing that we reject the present State, as we do any centralized State, for this role." Hugo Ball's thesis (#14) supports Landauer's view: "The German concept of State has destroyed German faith. It is the German concept of the State that I am targeting in this book." Therefore, "Following this idea to its logical conclusion brings us back to our cooperative socialism in which the individual communities and cooperatives in the city and on the land would be the heirs of their members."

Landauer believes that the great majority of workers and members of the various socialist parties have "an honest desire for socialism." But momentum is lacking because there is no general agreement about the agenda of "socialism". He therefore applauds the agenda and viewpoint of *A Socialist Program,* presented by Max Gerber et al. (#89). The authors specify "no new party—opposition rather to all parties; no new dogmas—rather...the education of independent individuals." Thus this socialist program is founded on the certainty of moral values. The authors identify three socialistic "weapons": propagandizing the idea, action (cooperative movement and education system), and the strike. "It is cooperative maintenance through cooperative labor." "From this follows the rejection of the contemporary State as well as all forms of State socialism." "The new society will be underpinned by the principle of federalism; the family and the community are its units. The ultimate form in which the economic body is articulated would be a system of councils." In summary "...this socialist program rests on two basic principles: on the one hand the socialization of the economy by cooperatives, and on the other, the socialization of the society by the creation of free and self-confident individuals."

# CHAPTER FOUR

## Justice

Landauer's concept of justice is omnipresent and pervasive. He watched the events of the revolution with care and concern, even suspicion. He was not in any repect an "insider"; as events unfolded they were judged against the standard of his own principles. From his Red Cross assignment in Poland, Landauer could not have been in a position to evaluate the rapidly occurring, unconnected events in Kiel, Munich and Berlin. He followed with keen interest published reports of the creation of Workers' and Soldiers' Councils throughout Germany, of the abdication of Wilhelm II, the signing of the Armistice, and the unplanned announcement of the Republic by Philipp Scheidemann and the acceptance of the presidency by Friedrich Ebert. The pace of change was indeed revolutionary, and the succeeding events orderly, if only hastily planned.

Friedrich Ebert began at once the transformation of the Imperial to a republican government: new personnel, new organization, new plans. Hugo Preuss, a distinguished academic scholar of constitutional law, was asked to draft a Constitution for the republic; January 19, 1919, was set as the day for election of a National Assembly, among whose duties would be perfection and ultimately adoption of the Constitution. To the Ebert government fell the task of providing continuity in a time of chaos. The monarchy must now be replaced by a democratically oriented legislature, judiciary and executive branch. The war was ended; thousands of soldiers were returning to Germany daily; the allies' blockade limited the urgently needed food supply, and hunger was common. Absorption of the military officers into a defense organization made them available to carry out the missions of the not yet approved republic. Continuing emergencies, impatience of the beleaguered populace and delayed or limited solutions to recurring problems— unemployment, hunger, inadequate housing, uncoordinated transportation and communication systems, continuing response to inevitable and unsatisfactory proposals for peace—a variety of internal affairs, offered opportunity for military intervention in civilian problems. Such action created increased tension among the people and greater suspicion of the new and democratic *regime*. An undercurrent of support for return of the monarchy actively opposed success of the republic.

The constitution was the subject of public debate for months, and it was adopted July 31, 1919. The Preamble sets forth the objectives clearly: "The German people, united in all their racial elements and inspired by the will to renew and strengthen their Reich in liberty and justice, to preserve peace at home and abroad, and to foster social progress, have established the following constitution:" (The scope and intent are suggested by the chapter headings.) "Chapter I: Structure and Functions of the Reich. Chapter II: Fundamental Rights and Duties of the Germans." There are 108 articles in Chapter I, 72 in Chapter II. The Constitution became effective August 11, 1919. It was, to be sure, a model of democratic governance; and significantly, as a codification of social and political behavior and the mechanisms to achieve and maintain it, it bore the stamp of approval of the National Assembly representing a wide range of German opinion.

Landauer reported (#102) a volume by Hugo Preuss in which "the Constitution is very well explained," and he notes that the text of the Constitution is reproduced as an appendix. But he does not praise or analyze this ambitious document; on the contrary he concentrated on the ways in which actual practice fell short of the utopian goals. Landauer was distressed that a central government with the guidance of a carefully conceived constitution could find it possible to disregard its own rules.

For example, at the moment when deliberations on details of the Constitution were approaching finality (June, 1919) publication of the *Republic*, a newspaper of the German Workers' Councils, was banned for "an indeterminate period" (#13, October, 1919), and Gustav Noske on the floor of the National Assembly had attacked Wilhelm Herzog and Ernst Däumig, editors of the *Republic*, "who now want to cut political deals." To Landauer it is ironic that the new Constitution by which many proclaim Germany the "freest democracy in the world," expressly forbidding censorship, has evoked no corrective measure, even months after this extraordinary governmental action. The *Republic* was forced into bankruptcy and ceased publication. Landauer summarizes: "The National Assembly falls silent, and then talks on—so far revealing as little of its democratic as of its Christian heart (No. 13)."

The incident portrays the administrative problems of the Weimar government, with occasional or frequent reversion to dictatorial measures. Because suppression of the *Republic* was a political act, political problems appear foremost, but this was a time of extraordinary difficulties in economics and trade, of returning soldiers and unemployment, of burdensome financial problems, of social unrest, of design and management of transportation and communication; it's a long list. Landauer's personal concern for justice, individual rights, freedom of expression and his abhorrence of violence serve to highlight the government's period of adjustment. Foremost among evils of personal transgression is crime.

Landauer commented (#26, October, 1920) "The *Republic* so far brought us nothing but an intensified Prussianization—an enforced economy together with moral anarchy—but what we need is self-government through cooperatives, in towns and districts."

## Crime

He attacks particularly political crime. "The death penalty does not only apply to vile crimes. Unwelcome beliefs, contrary political convictions suffice to extinguish human life."

Landauer had become well acquainted with Dr. Emil Julius Gumbel, a *privatdozent* on the faculty of the University of Heidelberg. He had taught Landauer the elements of statistics, and they had published an analysis of a portion of the data from Landauer's doctoral thesis. During these years which were critical to Landauer's academic progress, Gumbel was devoted not only to his profession but also to governmental affairs (Jansen, 1991). From the time that the Republic had been declared, he had followed closely the ruthless tactics of the Weimar leaders. Impatient for a smooth and orderly transition from monarchy to republic, they had become over-zealous; they not only condoned political murder, they freely accepted, and perhaps sought, the help of organized reactionaries. Gumbel had painstakingly assembled the data for a crushing revelation of systematic murder authorized by the Weimar government in collaboration with the judiciary throughout Germany. This information had been published in 1924 as *Four years of Political Murder.* Landauer's review of the book had appeared in the August 1, 1924 issue of *Die Menschheit* (#114). This review was not unique; he had published by this time more than a hundred of his political papers. The review and the Gumbel Affair, described below, are clearly coincidental, linked only by the participation of Landauer. His review begins, "There have been political conspiracies in all times....Still, the history of recent years has distinguished itself above all other periods by the conspiracies which virtually characterize its political confusion. In vain one seeks in the history of other peoples a parallel to a nation decisively defeated in an unprecedentedly savage war of conquest that is still sympathetic to the powers and military leaders responsible for it; that a people which, after all, has to its credit not inconsiderable intellectual achievements and likes to think of itself as a leader in culture, falls prey to the most nefarious demagoguery simply because it is blinkered enough to believe that the economic collapse caused by the war would disappear with a new *regime*; that a nation, without passion or even character, altogether ignores the judgment passed upon it simply because its own sense of justice has been lost—and finally, that the schemes of reactionary conspirators are not impeded simply because the people themselves do not see the immediate consequences of their behavior for international relations."

Gumbel tabulated hundreds of instances: by name of victim, profession, method of execution, person responsible for the execution (when known), name of the executioner (when known), and fate of the person responsible and of the executioner. Many other details are given, such as the number of murders committed by the right (354) and by the left (22). The average sentence of those on the right was four months; of those on the left, 15 years. The Ministry of Justice was charged with reporting Gumbel's disclosures to the Reichstag; however, its report was delayed for three years. Landauer pointed out in comments on Gumbel's related book, *Conspirators,* that "this organized political murder would not be possible without widespread, officially favored organizations". Landauer applauds the "documented account of these murder centers from which we learn about their actions, their interconnections, their objectives, and as their principal character, their terrifying mindlessness."

Walter H. Kaufman (*Monarchism in the Weimar Republic*) lists more than a hundred "militant leagues and secret organizations". Among these was Organization Consul (O.C.), whose name was later changed to New German League; Kaufmann quotes, "There may be cases where determined men...are needed to do away inconspicuously with political opponents. One must kill such dogs with a bullet or poison; anything else, however, will do as well." Kaufmann adds, "In the face of such language Gumbel seems justified in charging that 'there was during those years no political assassination in Germany in which the O.C. directly or indirectly did not participate'".

## The Gumbel Affair

On July 26, 1924 Gumbel convened an anti-war rally in Heidelberg's Town Hall. He was joined on the rostrum by Prof. Balmain of the Sorbonne

„Herr Gumbel, so fällt man auf dem Felde der Unehre!"

**"Mr. Gumbel, that's the way one falls in the field of dishonor"**
**(Kladderadatsch, Berlin, Sept. 11, 1932)**

and Member of the Baden Parliament, H. Stroebel. Gumbel concluded the rally with a request that attendees rise and maintain two minutes of silence in memory of those who had been killed in the war. He then added, in closing the meeting, that the deceased, "who I do not wish to say have fallen on the field of dishonor, have, however, found a horrible death". At once the students in the audience petitioned the University to dismiss Gumbel. He was then suspended, by order of the Honorable Willi Hellpach, Minister of Culture, Baden.

Landauer responded quickly, first in a complete analysis for *Die Menschheit* (August 15,#115) he says, "Among German Universities Heidelberg counts as the most progressive. How outdated that is, the University Senate has now documented by its conduct against its lecturer Dr. Gumbel." He then defends Gumbel by logic and by comparison of his "crime" to the more blatant "obstruction of orders of the government against the republic" by Professor Philipp Lenard, a Nobel laureate in physics. Secondly, in an appeal published in the Heilbronn *Sonntags-Zeitung*: "We call on everybody, but especially on all those who have been students or members of the University of Heidelberg, to lodge their protests to the University" (#116, August 17). A short time later (August 19) the Heidelberg *Volkszeitung* editorially said "An eerie silence has descended on the Gumbel case. No one knows whether this is a good sign, or if there is something ominous in it. The actions of the competent university authorities proceeding, as one is accustomed, at a slow—but surely very "thorough"—pace, have again come to a halt. We hope that the decision is very soon reached. Of great interest is a letter sent to the University Senate by a member of the University about the Gumbel case."

Landauer writes (#117, August 19): "Having been a member of the University of Heidelberg as an assistant until August 1, I feel I have the obligation to protest against procedures that—as it seems to me—must impair the reputation of the University most seriously in the eyes of any unprejudiced thinker within, as well as beyond the borders of Germany. The action taken against lecturer Dr. E.J. Gumbel remains outrageous even if seen from the perspective of utterly opposed political views. The only issue here is justice." "Now, it seems to me, the task and duty of the University is to remedy as quickly as possible the mistakes it has made.....You should be concerned that the University of Heidelberg does not present to the world the ignominious spectacle of intellectual slavery or the suppression on any pretext of persons who are a thorn in the side of enemies of the republic.

"If Germany wishes once again to fight its way back into esteem among the comity of nations, that can occur only by unremitting intellectual honesty, by a conscious sense of responsibility for justice and tolerance towards all the honest intentions of those with clean hands. In the case of Dr. Gumbel the University of Heidelberg faces decisions in accordance with which its moral credit or discredit will be adjudged!"

Curt Herbst was not only a distinguished scientist and Landauer's mentor; he was also Dean of the faculty and Director of Heidelberg's Institute of Zoology. Landauer's appeals through the Sunday newspapers had been published without prior consultation with Herbst or even notification. A colleague and close friend of Landauer, Dr. Karl Baldus, wrote to Landauer on August 9, "I am very sorry that, after reviewing your letter, that we are still at odds. I can tell you again that Herbst does not agree with your action, and he would like to see if you will withdraw the publication." He continues, "You know nothing about the recent course of the matter, and nevertheless you affect tones in your Senate letter which are neither necessary nor correct." Baldus explains that Landauer's identification of himself as a retired assistant implicates the supervisor of the assistants, Herbst, in consequence of which Herbst will be taunted. "You are behaving, as Herbst says, exactly like Wilhelm II: impulsive, rash and, above all, obsessed with what you are doing." "In any event, I tell you my opinion just as it is, not in such a way as to please you."

Clearly Baldus had talked, perhaps several times, with Herbst. On August 18, Herbst wrote to Landauer, "I am not the least bit cross with you because you interceded on behalf of your friend Gumbel, and I cannot consider your action unfair. Only because of its tone do I regret that you wrote the letter for publication ..."

Landauer had been away and did not answer Herbst until August 29. (We quote at length from this letter because it not only responds to the perceived impropriety of publication, but, of even more significance, clearly sets forth the principles which guided Landauer in this and many other instances. He said, "I am extremely sorry that...I have incurred your displeasure, and indeed, to a very high degree. It is not my intention to defend myself, but please allow me to point out two things: first, I must conclude from your first letter that you know the essentials of the *corpus delicti*, and then you already knew at that time that I had asked that my telephoned pledge concerning the withholding of publication be considered undone. Dr. Baldus told me at that time that in no wise was there an intention to bind myself in this regard. I must further assume that you knew from Dr. Baldus that I left the decision, whether or not the letter should appear, in Dr. Gumbel's hands.

"Of course I knew that publication of the letter offended you, and the great debt of gratitude that I owe you perhaps required that I accede to your wishes without question. The thousand-fold debt of thanks was constantly on my mind, and it will remain so in the future. In order to escape from the dilemma...I considered it most correct to allow Dr. Gumbel to make the decision, for I told myself that, even though I had interceded for him, the decision must remain with him..." Landauer had clearly and unequivocally expressed his respect for Herbst and his genuine and sincere debt of gratitude. The content of his letter to the Senate and its publication hinged on his agreement with Gumbel, a personal, not an institutional, matter. Landauer now presents in the same letter his position. "I know something about the 'truths' of human opinions and institutions, but I still believe that there are certain norms which govern one's conduct. I even believe that in emotional opinions, involving a sense of justice, it is much easier to reach agreement than in practice because in these very decisions personal constraints come into play.

"I have often posed the question to many different kinds of people in one form or another whether they think it is right that people these days should be in effect starving to death. I have not yet found anybody who had the courage to say yes. Nevertheless it is a purely emotional judgment. In practice people will find a thousand reasons to show that this is a necessary evil. Since I have seen on the one hand an emotional agreement in their judgment, on the other hand I assuredly believe that there are ways of preventing this evil, my own conduct is unequivocally determined. In this case it means a struggle against capitalism. I declare war on this institution, and whenever I find it possible, I point out the baseness and injustice of its conduct. But just because I claim that middleman's trade is eminently unjust I do not say that all tradesmen are dishonorable and scoundrels. You understand that I cite this example because it is closest to me, because my father is a tradesman and because I know very well his concurring opinion and feeling for justice. The same applies when I attack other institutions or processes, whether it be the church, a party, war or the injustice of a university. I believe that you understand what I am saying.

"It follows from this, of course, that I regret deeply the manner in which Dr. Gumbel has retreated. There are indeed lots of people who maintain that wars are necessary, and I demur at saying that they assert this against their better judgment, but I have never encountered anyone who could advance moral arguments for human slaughter that could persuade anyone else. I therefore find the field of honor no less dishonorable than capitalism or anything else. That here and there there are very many honorable, even absolutely pure people, requires no discussion. With the best will in the world I am unable to make myself understand sincerely how anyone's feelings could be hurt by the expression 'field of dishonor'.

"But I really do not want to burden you any longer with my arguments. I know that you have reason to scold, and all that remains to me is the possibility of regretting this most deeply. I shall always remain yours gratefully, respectfully and truly devotedly."

Herbst had genuine respect and affection for Landauer. They had been in a very close professor-to-student relationship for four years. He respected

Landauer's right to help Gumbel, even to publish his opinion in an appropriate place. But he was upset by the tone, by the manner, by the very fact of communication with the University Senate directly. He closed his letter of August 18, "Be assured that, in spite of the Gumbel affair, you remain dear and respected, and I will follow your future progress always with great interest". On August 22, however, he says: "Only yesterday after the regrettable publication of your letter to the Senate did I learn fully of the content, and I must unfortunately tell you that I was extremely annoyed and shocked at the manner in which you wrote the letter. There appears from your letter such a self-righteous arrogance that I wonder how you can have almost no sensitivity about it, so that you could draft and even publish the letter. I have held you in high esteem as a serious scientist and as reliable, and it worries me now that you turn out to be a raging fanatic who belongs in the same category as the students who sent the other arrogant letter to the Senate.

"I do not hold it against you that you interceded on behalf of Gumbel; indeed, I find it praiseworthy that you befriended him, but I do not like at all the moral arrogance of your decision on this question, based entirely on law and therefore relative and indeterminate."

Finally, on September 12, Herbst says, "After your and my last letters the issue is resolved. That's why today I want to send only cordial best wishes. Have a good trip, and may you be blessed with lots of success over there. However, get accustomed to expressing your antipathies in a different tone, and only then can you succeed in something, or else you will disappoint the world even in America, as already seems to have been done in Germany.

"Best wishes and accept my greetings a thousand times over."

Herbst and Landauer corresponded irregularly through the years, mostly by way of exchange of reprints. On his death bed, March 30, 1946, Herbst dictated, "Dear pupil and friend: We have worked together so closely and productively that you have become to me one of the dearest men on earth. Only Theodor Hinderer, who died in the first world war approached you, and together laid the groundwork for the role of the nuclear material more firmly than my original research had. I have set this forth in my last paper on theoretical biology, and I want you to have it as soon as it is typed—and then to resume our personal correspondence. I still enjoy very much receiving reprints of your scientific work."

## The Judicial System

"If the reputation of the public prosecutor was not high even in earlier times—among the people as little as among jurists—the revolution and the jurisprudential processes connected with it have completely undermined public confidence in the office" (#12).

Replacement of the military court in Bavaria by the "People's Court" offers no real advantage. Power, in trials affecting life and death, rests in a single hand, "and the accused is deprived of the possibility of appeal." Landauer finds it appropriate to quote Alexander Herzen: "For them the judicial process is only a cruel game, like bullfighting for the Spaniard. The public prosecutor feels as humiliated and insulted as a skillful toreador, if the tormented animal is alive at the end" (#12).

Landauer finds the words of Thoreau *à propos*: "The law has never made men more just; on the contrary, good men fall prey to injustice because of respect for the law." He suggests that an approximation to real justice may result when laymen and professional jurists work together, as in the Ledebour trial.

## Welfare

Landauer made it clear that all of society, indeed humanity, should benefit from the combined talents of all its members. He repeatedly asserted his recognition of the importance of the talent and potential of the individual. The sum of individuals' contributions was perceived as an indispensable dynamic force. The size and diversity of societal groups, coupled with failure to understand the value of every individual, inhibited the design of a plan that would assure maximal individual productivity. Some portion of any population was incapacitated briefly, seasonally, perhaps for a lifetime. Temporary failure of the individual to contribute because of illness, ineptitude, or inability to cope with uncertainty, pointed to the need of a readily accessible safety net.

The need arose with such regularity that it became incumbent upon society to invent a solution. Creation of a system to protect the infirm or unfortunate was contingent upon the principle of equality, the sanctity of the individual. It was a question of human rights. And even this premise required acknowledgment of the unpredictability of disease or deformity, of economic conditions, or suitable opportunity. The complexity of interactions within a group or between groups, e.g., nations, required that each group include human rights among its principles of conduct. The basic premise must be respect for the individual.

Because the problem seemed to be so easily identified and the solution so elusive, Landauer was intrigued by the ideas of Josef Popper-Lynkeus a Viennese engineer who turned his attention in later life to social problems. He identified the "social problem" as "the problem of finding an institution suited to assure each individual of the necessary economic means to sustain life, without making this dependent on the will of other people." This definition was restated as "the recognition of the equal value of all individuals as concerns their need to be assured of their economic necessities."

Popper originally announced the principles of his welfare program in 1878 on the occasion of the 100th anniversary of Voltaire's death: *Das Recht zu leben und die Pflicht zu sterben* (first edition, L.Reymann Verlag, Leipzig; 4th edition, 1924, R. Loewit Verlag, Vienna). Wachtel has reviewed the content of this interesting volume. Popper announced with great conviction the need for a program of minimal subsistence for all.

He asks "How will this minimum be provided? By the introduction of general subsistence which we propose to establish through general conscription as a means of unconditional distribution of a living minimum in natura." "We want to recognize the right of every man to a worry-free physical existence, just as we recognize the right and protection from the enemy by a military force. War is only an extreme example; the battle against need is a regular occurrence, all the more reason that we want to defend ourselves against this utterly persistent enemy." "We safeguard every citizen by this army without regard to status, sex. age, religion or capabilities; we want to do exactly the same regarding the protection of the destitute."

Popper reiterates his concept of the sanctity of the individual in several later volumes:

1905. Fundamentals of a New Political Constitution
1910. The Individual and the Evaluation of Human Existences
1912. Universal Subsistence as the Solution of the Social Problem
1916. Autobiography

Landauer has summarized Popper's program in a series of essays (#104, #105, 106 and #124). The details of Popper's plan are faithfully described, together with Landauer's evaluation. The fundamental purpose of the plan is to guarantee subsistence ("not only food, housing, and furnishings, clothing, heat and light, but also medical assistance, nursing, medications etc.") to every citizen. In addition to these primary needs, "there are also secondary or cultural needs (books, theater, travel etc.)." The program would be managed by "The Ministry of Living Standards". The principal mechanism for this program is "The Labor Corps", a segment of the population pledged to serve a term of several years, less for women than men. The Labor Corps would be called upon to produce and distribute all materials necessary for the program. The category of secondary requirements, because they are so varied, would not be provided as real goods, but rather as cash, "to be raised out of the proceeds of state monopolies, levies, tax moneys, etc." (#104). A significant part of the plan is the coexistence of a "free private economy" which would apply only to non-essential goods, a "luxury economy". "This would in large measure avoid at least the currently disastrous consequences of a capitalist economy". Landauer notes Popper's reasons for maintenance of the "minimum organization" and also his reasoning in the preservation of "a free private economy".

Landauer (#105) reports that Popper provides in a 200-page section a statistical analysis of the size of the Labor Corps, men and women, necessary to perform all its necessary services. He accepts the estimate of 7.75 million men serving for 13 years, and 5 million women for 8 years, working not more than 7 hours per day. But he takes issue with the administrative centralization and its implied parliamentarianism. Conversely he is disappointed with Popper's conception of the role of community in the plan, preferring "a constitution the binding principle of which is a federation of unions of self-governing communities."

Fritz Wittels (1880-1950) in his memoirs (edited by Edward Timm, 1995) acknowledges the work of "his good friend, Josef Popper," and issued his own program (1922 in German, and 1924 in English), ambitiously titled *The End of Poverty.*

In a separate piece (#108) Landauer summarized the three detailed essays on the Popper plan. He notes that this "revolutionary" plan is based on ethical principles discussed in Popper's "The Individual" and other serious essays noted above. The success of the plan is dependent on a general acceptance of the inherent importance of the individual and the willingness of the general population to carry out the plan. It is an ambitious plan which even Popper conceded (1912) could not be put in place before 1980.

## The Role of Cooperatives

Landauer's proposal of a nationwide system of cooperatives was made basically upon principle, but it was supported by his study of operating communes. The diversity and adaptability of the cooperative enterprise in Germany encouraged him to think seriously about the feasibility of systematically extending the plan.

Landauer's interest in collectives, *Siedlungen*, was signalled early. Already in October, 1920 he had formulated opinions about such enterprises. He reflected (#25) that "The capitalist economy has bred its own downfall." Highly intense competition has led to insufferable mistrust among people, and indeed among nations. "Only this mind-set could have led to the World War..." "So far the revolution ... has proved sterile throughout the squabbles of the parties in the struggle for governmental power."

He took the occasion of the founding of a new periodical to reaffirm his position. The *Bauernzeitung*, although aimed at farmers, "is also important to all people who take part in building the economy of Germany and of the whole world." "The only future for farming and essentially for humanity lies in a general working community which is built not upon envy, hatred and competition, but on mutual assistance and love—we have in mind cooperatives."

This idealistic scheme is conceived to start with the commune and to progress from village to district organizations as a link between city and country. "Middlemen, the actual abscess of capitalism, must disappear." "...the spirit of joint work and mutual assistance must be established. That is the school". Landauer explains that the *Bauernzeitung* will provide technical information to farmers while offering continuously advice on the virtues of communal living. If properly managed, profits from the commune will be reinvested in the community, particularly in education. He suggests for a motto, "The premise for a free humanity is the cooperative free peasantry."

## The *Siedlungen*

Reviews of active settlements are presented with the objective of demonstrating the feasibility of Landauer's concept of the commune as a national enterprise. "...It will be the purpose of these writings to sketch, with critical interpolations, a picture of the German settlements so as to illustrate the efforts within the settlement movement that could be of significance to us."

The net effect of these reports is a description of a new Germany, one in which the capitalistic system is replaced and the central government is superseded by a nationwide federation of cooperatives. Such a program has two principal premises: access to land, whether privately or publicly held, for commune development, and large scale participation of the people in the enterprise.

The magnitude of the task is suggested by a few observations. The German population in the early '20s was of the order of 70,000,000. The country was faced with enormous reparations payments, while attempting to rebuild its agricultural industries and redirect its manufacturing industries in the interest of a people at peace. Politically the country had experienced revolutionary change—demise of the monarchy, creation of the Weimar Republic and a new constitution which prescribed a representational government, with an elected president, a parliamentary legislative body and a judicial system. The national economy required the installation of a system of self-sufficiency, investment of foreign capital and the rebuilding of foreign markets for export. The government was challenged at every turn to extricate the country from social and economic chaos. The temper of the times was not conducive to a nationwide socio-political experiment.

Nevertheless, Landauer wrote descriptions of 6 settlements. They show similarities in organization and purpose but differences in emphasis according to geographic location. The six settlements are: (1) Bergfried Settlement Cooperative, Munich (#43); (2) Habertsdorf, Kassel (#45); (3) German Workers' Union for the Promotion of Working and Settlement Cooperatives on the Model of Voelpke (*Deutscher Arbeitersbund*), Magdeburg (#46); (4) Empor, near Leegebruch (#49); (5) Orchard Cooperative Eden, Oranienburg, near Berlin (#51); (6) Intensive Settler School, Worpswede (#53)

Similarities among these settlements are:

(1) Incorporation of Christian principles
(2) Available land (19-625 hectares=47-1500 acres per settlement)
(3) Sources of membership and initial investment
(4) Privileges of tenants, e.g., housing, garden, exchange of goods within the settlement
(5) A system of management and commune promotion, including exchange within the settlement and export of surplus goods
(6) Educational plan
(7) Self-insurance pool
(8) Welfare plan, e.g., orphans, sickness
(9) Year-end sharing of proceeds

Specialties which were developed among the various settlements include fruit farming (Eden) (#51), crafts and skilled trades (all, to some degree), coal and potash mining (Voelpke) (#46), producing and productive associations (Empor) (#49), and intensive land use for agriculture (Intensive Settler School, Incorporated) (#53)

Having presented rather detailed statements about several settlements with interpolated comments on others, Landauer provides a review of his position not only on the settlements discussed but on the prospects of the settlement movement (#59). Because of the comprehensive nature of this essay, it is reproduced in its entirety. It elaborates Landauer's views on the settlements in relation to the principles of socialism and capitalism.

# CHAPTER FIVE

## Epilogue

Landauer began his new job, Assistant Geneticist in the Storrs Agricultural Experiment Station, October 1, 1924. Working with his new colleague and mentor, L. C. Dunn, he was industrious and successful in the new and difficult field of poultry genetics. He published regularly and reported his results to international congresses of genetics, poultry science and developmental biology.

The Gumbel Affair was fresh in his mind, and he was continuously concerned with the events which led to the demise of the Republic and the sweeping success of the National Socialist Party. The writing of his political essays had required close attention to national affairs in Germany, which perennially conflicted with his strong belief in the importance of the individual, his opposition to violence, and his opinions about capitalism. He was deeply concerned about the application of the principles of Communism.

As he carried out a rigorous curriculum of studies in preparation for a professional career in biology, Landauer was apprehensive, even bitter, about the behavior and prospects of the German nation. He saw fumbling, cruel and callous mishandling of human resources; moral disintegration of an entire people; neglect of the welfare of the masses; wasted human potential; ineptitude and deceit in a government that had been given wide constitutional latitude in the new republic. He observed many faults in the structure and operation of the Federal Government; he was bold and unflinching in his analyses and criticism.

At the center of the Gumbel Affair was the principle of freedom of speech, which Landauer prized highly. It presented itself again in the "incident" on the Storrs campus. The Board of Trustees of Connecticut State College adopted a resolution April 17, 1935: "that any formal public agitations or formal public discussions on the campus by...the college staff or individual students...will subject such individuals to cause for removal..." Walter Landauer was at the center of the community response; within days he had sent letters to about 40 people, including "all American winners of the Nobel Prize" ( Landauer Collection, Dodd Research Center, University of Connecticut Libraries). He later wrote to L.C. Dunn : "I have not been able to work for weeks" (Landauer-Dunn correspondence, Archives, American Philosophical Society, Philadelphia). In June, President McCracken resigned, and search for a successor was begun. A candidate was chosen, Albert N. Jorgensen, and the Board of Trustees at its October meeting authorized the following statement: "It appears to the Board that the conditions have been removed which seemed to make necessary its resolution of April 17, 1935 concerning subversive activities and military training at the college. The Board, therefore, withdraws said resolution..."

Landauer's involvement in German affairs during the early years of the Weimar Republic through independent essays and his book reviews and, prior to that, his War service with the International Red Cross, had left scars that never healed. The depth of his feeling is revealed in three letters, two written in 1946 and one in 1960.

In response to a letter in 1960 seeking support for the celebration of the 600th anniversary of the University of Heidelberg, he wrote: "When I came to Heidelberg in 1919—following that famous revolution—one of the first things I learned was that Professor Otto Buetschli, the celebrated biologist under whom I intended to work, was *persona non grata* in the academic community because he had entertained some reservations about the German brand of martial spirit during the first world war. Moreover, at about that time also Professor Lenard, the Nobel Prize winning physicist, drove from the university a friend of mine by the simple expedient of not accepting her for the physics course required of medical students, because she was Russian; the university did not protect the rights of the student although requested to do so. During the ensuing years I witnessed the—ultimately successful—campaign to oust Privatdozent E.J. Gumbel. The savage behavior of the university authorities did by 1924 succeed in convincing me that there was no academic future at Heidelberg for one with my name and unorthodox political convictions. Hence I resigned from my assistant position.

"If the University of Heidelberg were now to collect funds for a memorial to the many colleagues who were driven from their posts during the thirties, I should be happy to contribute liberally; but far from it, those people have been treated most shabbily....Why should we, who were despised and unwanted, be expected to honor the University? Let the University come forward and honor the victims of its recent past. A start in that direction might well be made with Prof. E.J. Gumbel, now at Columbia University, who was the first to be driven out."

In the summer of 1946, Landauer's response to a letter from Prof. W.E. Ankel about "international scientific relations" touches on other issues whose origin in the early '20s helped to shape his opinions and to precipitate later action. "I foresaw clearly as early as 1924 that the German republic was doomed since the government was too weak to protect its own officials from murder and since there was not enough sense of justice among the people to impose an improvement of what had already then become an atmosphere of intolerable unfairness. Having partly Jewish ancestry and counting relatives and friends among those murdered and prosecuted forced me, without doubt, to realize more clearly what was going on. I cannot admit, however, that the events which followed and which culminated in the usurpation of power by the NSDAP left any doubt as to the position which persons with a sense of justice and fairness should have taken.

"It is with such persons only that I care to renew friendly relations. I say this with some reluctance and much sadness, but it is so. I have no feelings of vindictiveness, nor do I think that I am hypocritical in this attitude. The one experience that is certain to remain with me to the end of my life is that of the horrors of deliberate torture, persecution and assassination of millions of Jews and liberals in Germany and German-occupied Europe. Even one's enemies must understand that one cannot erase these memories from one's mind. Nor can one forget that these events failed to arouse indignation and protest among the German people."

Landauer, after award of his doctoral degree, had been appointed by Professor Herbst Assistant in the Zoological Institute of Heidelberg; Dr. Karl Baldus was a close personal friend and colleague, to whom Landauer voluntarily yielded half of his Assistantship. Baldus died at an early age, and Landauer dedicated a 1927 paper (*Roux Archiv*, 110:195) on chondrodystrophy to the memory of Karl Baldus. Also in 1946, Greta Baldus wrote Landauer in an attempt to renew the Heidelberg friendship of the early '20s. Landauer's reply was brutal and revealed the depth of his anti-German feeling, extending even to the personal relationship with the widow of a very dear friend.

"During the war I was told by a mutual acquaintance of ours that after 1933 you had adapted yourself rather well to the Nazi *regime* and had broken relations with former friends who were not personae gratae with the Nazis. Since you also ceased writing to me long before the war cut off international connections, I had little reason to doubt the correctness of this report.

"I am not ready to believe that you had real sympathy with the Nazis. Your attitude was perhaps determined by motives of fear and self-preservation. Even in that case, however, I cannot admit that everything is now as it was in the good old times. You know that my friendship for your husband was one of the great events of my life. I feel certain that he would not have compromised himself. Perhaps I should be willing to forgive a great deal for the sake of his memory. However, too much has happened in the years preceding the war and during it, ever again to forget.

"You inquire about my family. Well, my mother died in Heidelberg in 1943, in loneliness and isolation after one of her children had been killed by the Nazis and the others were scattered over all parts of the world. If any of my pure-blooded Aryan friends paid her as much as a single visit I did not hear about it. Any such sign of loyalty toward me might have taken a little moral courage (really not too much of it), but would have meant and would mean now a great deal to me.

"One might get over the merely personal tragedies, but I am afraid that it will be impossible ever to forget the torturing and murder of millions of Jews and anti-Nazis, nor the way in which most people seemed to adjust to it.

"If there were real signs of moral regeneration now, my attitude might be different. There is very little indication of it. All that I hear is that nobody in Germany now admits that he ever had any dealings with the Nazis. There seem to be frantic efforts to trim one's sails to the American breeze—without any convictions about it. It is not a very inspiring spectacle.

"I could wish that your first letter had at least not included a request for food. Without this material motive of your letter, I could have believed more easily that you were sincerely anxious to renew a friendship which you had discarded earlier. You must not forget that there are millions of refugees in Europe whose situation is infinitely worse than that of the German population. It makes me very sad to have to write this letter, yet I cannot see how I could have done better"..

Respect for the individual and the need to understand the forces which determine potential persisted throughout his career. He was sufficiently persuaded of the genetic basis for deviant behavior by Johannes Lange's study of identical and fraternal twins, that he reviewed the work for both German and American journals. He summarizes (#131): "The attitude of society toward the criminal as concerns social determinism is hypocritical and repulsive; and, as a result of these biological analyses, is ridiculous and outrageous. The fiction of the responsibility of the criminal has received its death-stroke. Responsibility of society to the criminal has become a documentable disgrace of our time."

This posture led him to an interesting professional relationship. In 1944, he was asked to review a new book, *Stroud's Digest of the Diseases of Birds*. Landauer, a specialist in fowl genetics, wrote: "I think that the book is extraordinarily well written...in simple, lucid, excellent English...One has the impression that the author approached his tasks without bias, with sound scientific reasoning, and with care to avoid possible errors." After pointing out certain "errors" and "blemishes", Landauer adds: "The author is unquestionably a very serious-minded and determined person...The composition of the book and many original observations are clear proof of good judgment and of very considerable powers of observation." The author of the book was Robert Stroud, a convicted and remorseless murderer whose death sentence had been commuted to life imprisonment by President Wilson. Altogether he had spent all his adult life, 35 years, in Federal prisons, most recently Alcatraz. He had performed research of a high order while in solitary confinement with limited help, but with the ample and respectful tolerance of a succession of prison wardens and guards at Leavenworth.

Correspondence was frequent and lively for many months. Landauer sent books, reprints of his own work and literature references; Stroud offered critical comments and presented rational and detailed research plans. (The correspondence and the book are housed in the Landauer archives, Dodd Center for Research, University of Connecticut.) But these plans were not carried out. Stroud was deprived of the "laboratory" which he had created in his Leavenworth cell, and further, he was not permitted even to revise his book.

Thomas A. Gaddis has provided a biographical summary of the legal and personal details in the life of *The Bird Man of Alcatraz*.

Although communication with Stroud ceased, Landauer acquired from prison and judicial authorities details of the official Stroud record. Meanwhile, sociologists, psychiatrists and other specialists explored the relationship between aggressive behavior and the XYY chromosomal configuration in humans. Landauer accumulated a large collection of these scientific references; he did not actively publish in this field, but almost certainly he suspected that Stroud's behavioral characteristics were similar to those commonly associated with the XYY karyotype. But he knew, too, that intelligence, talent and social behavior were complex conditions ultimately determined by the relationship of specific genes to their genetic and environmental background.

At the root of his understanding was the individual and his potential. Within the individual he saw infinite good, and for society infinite good in pooling of talents; and immeasureable loss to a society which failed to realize the aggregate benefits of the diverse potential of all, or worse, which systematically suppressed the natural endowment of even a portion of its members.

However idealistic and unrealistic many of Landauer's proposals seemed, and perhaps to some, still seem, it is worth recalling that over time notions of the "safety net" of societal responsibility for children, the aged and infirm, have been unquestioned assumptions of a more humane society. The debate continues on how far and how fast to proceed along these lines, but as absurd and unfeasible as these ideas were regarded on their first appearance, they have in fact been substantially implemented in most economically advanced countries.

## 1. HEIDELBERG PROFESSORS AND THE QUESTION OF GUILT

*Die freie Zeitung* (Bern) III (6). 1/18/19

Surely, every decent German, at home or abroad, would have to be ashamed—even if he had already acknowledged Germany's sole guilt when the war began, or only just now come around to it—to hear how German professors are blind to the interests of our fatherland. 76 Heidelberg professors and lecturers have recently submitted to the People's Government of Baden a statement in which they demand that the People's Government use its good offices with the Federal Government to bring to a halt the random circulation of onesided documents which "allegedly" publicly admit German guilt. And instead of this "the proofs of the overriding share of guilt" of our enemies be unequivocally demonstrated in public, and that the purely German character of the territories claimed by French lust for conquest be most emphatically proven to the world.

While we are pleased with Kurt Eisner's well-directed initiative, and look around most longingly for other men who like him will unreservedly rip off the veil masking the crimes of the German monarchy, these reactionary scholars, who should after all be our leaders, are instead dragging the German people into further misery. Who could still wonder, then, that the former enemies of Germany—whom we have occasion enough to be reconciled to and to whom in our inmost thoughts we should be grateful for the liberation they have brought us—should have doubts about the honor of the German people. Where are the voices of the German university professors which should be raised in protest against such criminal assaults on the future of the German people? Does the new Germany want to entrust the education of its youth to such elements?

W. L., Heidelberg.

## 2. MONISM AND POLITICS

*Mitteilungen des deutschen Monistenbundes*, IV (1,2), 2/1/19, pp. 14-15.

### A Word of Warning to the [Monist] League Leadership

Today the League faces a decision about its conduct in the future which will awaken lively interest in all the circles of its membership. In the last issue of the *Mitteilungen* the contradictions within the League came to light in the sharpest contrast, with Hermann Schell and H. Peus on one side, and Professor L. Fraenkel on the other. The continued existence of the League is at issue;

large circles of the membership will make their continued participation dependent on the future behavior of the League, and they will surely be right! In the first postwar years politics was unquestionably pursued in the League newspaper, politics with which most members plainly were in agreement, more or less. Let me recall Vielhaber's poems "I am the people" or "To England," M. von der Porten's poem "1914" which remind one of Vierordt's "Songs of Revenge". When Maurenbrecher once said in the *Monist Century* that England "lit the fuse under the powder keg on which it was sitting" this was only one of many such utterances that could be cited in quantity from the League publications of that period. Dr. E. Meyer opposed pacifism in the most vehement manner in his letter to Professor Wilhelm Ostwald. Try reading today the result of Otto Juliusberger's poll, "Europe under German Leadership!". Even in the death announcements politics was introduced, and plainly by the League leadership itself. Was it really the case that all the many members who fell victims of the war "*willingly* sacrificed their lives for Germany's fortunes and future"? Quite surely not! In March 1915 the unfortunately all too prematurely dead Otto Borngraeber first pointed out in clear language the strange posture of the League in his "Open Letter" to the German Monist League (DMB). With Professor Ostwald's rebuttal that monism could be kept alive only with a German victory, Borngraeber's flaming protest was disposed of. At the same time I also addressed the League as a convinced pacifist, urging them to desist from unilateral politics within the League, and to avoid as much as possible taking a position on the issues of the day inconsistent with the League's commitment to "scientific perspectives." It was understandable that my warning was not published at a time when pacifists were despised in Germany and virtually regarded as traitors to their country. After a long time I received a written communication to the effect that the official position was that the League would not commit itself politically in any direction. In itself that is quite correct. Why, then, was politics not completely banned in the newspaper's ambit? Why was its orientation so unequivocally a onesided nationalistic one? Wasn't the official position of the League reflected in all those essays just the same? It is not my intention today to provoke a quarrel about past statements of opinion. Klabund quite correctly says in the *"Neuer Zuericher Zeitung"* that the war besmirched everybody, and that today each of us has to answer before the international court of world-wide conscience. By now it is unconditionally necessary for the League leadership to decide the question whether the League is to serve the discussion of questions of world view, or those of party politics.

Walter Landauer, Karlsruhe

## 3. BLOOD SACRIFICE

*Die freie Zeitung* III (No. 37) 5/10/19

Words can hardly express what Eisner's intellectual community feels at the news of his death. A breach has been torn in Germany's political leadership which will not be repaired. One of the few upholders of real, not superficial but true, revolution has passed away with him.

Many who have lovingly followed the destiny of the German people will be overcome by the bitter feeling that *freedom of thought, of speech, and of action* have now definitively *disappeared from Germany's precincts,* that anyone who dares to think otherwise than the reactionary masses is doomed to death. *Reaction* is bloody in its deeds and seeks out Germany's finest children as its sacrifice. Only a few weeks passed since Karl Liebknecht and Rosa Luxemburg were sacrificed by a military rabble to the goals of reaction. The government has still done nothing to investigate these events, or to take steps to punish the guilty parties whose names are after all known to everyone. It does not want to do anything for it is glad indeed to have gotten so easily rid of two people who, on account of their courageous truthfulness and impulse for freedom, were undeterred by any obstruction, and who were a constant reminder of its own lies and shortcomings. One can *condemn* all violent measures, and despite this understand and forgive Liebknecht's and Rosa Luxemburg's principles of action! For it was just the irresponsible reactionary activity of the majority Social Democrat government that forced them into their plight—actually coerced them. Any compromise would have had to be an irreconcilable contradiction to the fidelity of their self-sacrificing convictions. Perhaps by now it has dawned on a large number of working Germans that they have—once again—set the foxes to guard the henhouse. This knowledge will come, it must come—we can only hope not entirely too late.

Everyone will instinctively think of Liebknecht on hearing of Eisner's death—yet how much separated these two heroes of freedom. No appraisal of Eisner's activities will be given here—anyone free of prejudice can easily do this for himself from his speech and action. But what one would like to shout at every German: do you understand that with Kurt Eisner one of the best of our people has been killed? Do you know that he was one of the few who, from the beginning of the World War saw through everything and that he exerted all his efforts to save his people from a bottomless abyss? Do you want to place yourself on the same level with his murderers and approve this bloody crime? Wake up, German people, preserve your freedom, give heed to your most sacred rights! Every individual is responsible! Come to your senses at last and—act!

Hardly anybody in Germany dares any longer seriously to deny that Liebknecht and Rosa Luxemburg fell victim to a militarily organized gang of murderers—it is different in the case of Eisner. Already in its first reports on the events in Munich the press attempted to convey an impression to its readers that this was the act of a fanatic *acting alone*. We know all too well the *influence of the press* over the German people. The assassin may or may not have been associated with a political party—but his deed was the *consequence of the reaction that stood behind him,* and it must and can be assessed only in this wise!

Eisner was one of the few who "knew the score"! He took for granted the need always to state openly and clearly to all the world what he thought and what he meant to do. *Kurt Eisner* wanted to be to the German people what *Wilson* wanted to be to all the nations. Both of them *upholders of the humane idea of a new and free epoch.* Truly, great similarity can be found between Eisner and Wilson.

Just a few days ago the whole world waited expectantly to hear what Eisner would have to say in Bern, and we breathed a sigh of relief when we heard that at least one man recognized that only by acknowledgement of guilt, by atonement, love and the will to constructive cooperation could the German people find its way in joyful hope to the uncertain and problematic future.

The assured consciousness of the aims with which Eisner assailed the Berlin reaction, with which he clearly demonstrated Germany's guilt for the war to our people—long known to the rest of the world—the labor that he contributed to the development of the new Bavarian State, had won him the hearts of all noble thinkers and will preserve for him an unforgettable memorial among his friends. Opposed to all acts of violence, whether from the left or the right, Eisner attempted with his election slogan "rather go left in unity than to the right," to create the foundation on which a new building could be erected. He strove to educate the German people to truth and candor, to a universal love, in order to make it worthy to join the comity of nations.

On November 30, Eisner said to the Bavarian soldiers' soviet: "I am afraid only of one thing, I am afraid of our own fear," and thereby himself acknowledged what threatened and threatens danger to every free spirit in Germany. Clinging to the old ways, fear of the new, of material and social losses stand in the way of any freedom-loving development in Germany. Even the German workers have not yet recognized that they, as Marx says, "have nothing to lose but their chains. They have a world to gain."

Let all friends of freedom and of the future of the German people; let all those who understood and honored Kurt Eisner's deeds and aims; let all fighters for a new era in the history of mankind bear ever in mind the words of Eisner in his honor, and as a constant stimulus to tireless effort:

*"Revolutionaries are afraid of nothing!"*

Walter Landauer, Heidelberg

## 4. "COUNCILS" AND "PARTY" OF INTELLECTUAL WORKERS

*Die freie Zeitung* III (No. 37) 5/10/19.
by Walter Landauer, Heidelberg

The revolution in Germany has given birth in nearly all the larger cities to organizations calling themselves "Councils of Intellectual Workers." Even at first sight one can see that the most varied efforts dominate within these councils. Some of them—represented above all through the newspaper *Der Geistesarbeiter* "The Intellectual Worker" edited in Munich by Richard Staudinger—aim at an association of all intellectual workers within a "Party of the intellectual workers of Germany." The concept of an intellectual worker is here—surely because of a shortage of real intellectuals—very broadly defined to include within it all State and municipal officials. Whether Mr. Staudinger is especially called and entitled to appear as the leader of intellectual workers is something we will not discuss here. *Sapienti sat!* [The wise will know.] Let Mr. Staudinger, if he feels the need, add yet another to the "all-too-many" newspapers—it can hardly be so bad as to be especially noticeable among the rest. But it is a matter of little concern to the general public how it expresses in the name of intellectual workers their "cultural affairs." Its principal efforts are directed to drawing intellectual workers away from all political parties into an independently organized protection of their interests. As though the most important task of Germany's intellectuals today were their consolidation into their own party! The question whether more intellect is to be found among officials or among workers needs only to be posed, but not to be answered, and that the lower officials are in fact proletarians is equally well known. Therefore, a new party of this kind lacks any sort of intrinsic justification.

Not every council of intellectual workers would in any case find its reason for existence in separation from the people. Others, like the "Council for Art and Culture in Baden" in Karlsruhe, or the "Association of Intellectual Workers" in Hamburg, aim at drawing the most significant personalities in the world of artists and scholars into collaborative activity. In artistic and, in particular, in pedagogical concerns they have shaped all sorts of no doubt valuable demands into "programs," and even if we have so far heard little about efforts to fulfil them, still they could nonetheless stimulate their further development.

While these councils sacrifice their viability to Staudinger's convenience by seeking to unify people of the most heterogeneous political viewpoints in a narrowly circumscribed program, the other councils are rendered ineffectual from the start by other factors. Even if in some cases their efforts really pursue ideal goals and can be taken seriously, there are still grave reservations against them. What one must keep in sight, especially in this time of revolution, is that it was the *German intellectuals* who were the most effective advocates *for the German war*, and that even now the majority of them stand as barriers to the development of a more liberated humanity. Those—and they are still all too few—whose minds have been changed by events should not be surprised or even embittered to find themselves mistrusted by everyone; with an honest effort in the right direction they will be trusted again. Apart from the fact that it still remains the most important task and the precondition of every effort to bring about the triumph of the truth about the causes of the war and who is guilty of the war—men like Grelling, Nicolai, Lichnowsky, and Mühlon are as despised in revolutionary Germany as they were before—yet it would appear that the path of the intellectuals must be altogether different from that of the councils of intellectual workers. Foreigners will not be impressed by these councils, and the German *proletariat* will be readily inclined—perhaps not always without justification—to treat them as new (and unwanted) spokesmen.

Whoever is convinced that the revolution has so far failed to achieve a single one of the demands that we must await and insist on today; whoever is convinced that the revolution will continue—despite all the resistance and hesitations until its tasks are fully completed—knows also that it must in future strike out along quite other paths.

What we above all require is a new attitude—and the new leaders will then soon be found. If everyone takes honest pains to find out the truth, if everyone makes the effort to give up his prejudices, and no one considers the revolution as a road to better his social standing, but as a road to socialism, then the acutely needed and heartily desirable equalization among the classes of the population will emerge of itself—but this can only be delayed by the efforts of the "councils of intellectual workers". Of course the intellectuals should not wait until the revolution presents them with faits accomplis—in many respects it is already too late for them—but all of them should be working for the realization of the highest ideals that lie close to our hearts. Everyone must be conscious of the enormous responsibility that falls upon us today, for it is now that the fate of many generations of Germans—perhaps of the world—is being decided! Establishing "programs" is not our task today, but engaging the hearts of all men with great ideas.

It cannot be the task of the intellectuals to breed a new spirit of provincialism in order to impress the population from the top down—that approach is by now definitively over! Whoever now wants to achieve great things can do so only as part of the people, with the people, in order to realize the watchword: "We want to be a single people of brothers."

Walter Landauer, Heidelberg

## 5. RESULTS OF THE GERMAN REVOLUTION

*Die freie Zeitung* (Bern) III (49) 6/21/19
by Walter Landauer, Heidelberg

Whither Germany is now the fearful question which burdens the souls of all feeling individuals. Anyone disposed towards freedom—regardless of which nation—will at once feel deep and most bitter pain and compassion for this people which will permit nothing to delay its insane course into disaster and ruin. Whoever finds himself beyond the borders of Germany might judge a great deal more calmly, perhaps even more objectively—nonetheless he will never be able to share the oppressive sense of the total collapse of all morality, of a moral uncertainty bordering on characterlessness of the entire German people as it weighs each of us down, threatening to rob us of the last spark of hope.

The mentality of the German revolutionary leaders, the governing socialists, is too well known to require any commentary. We know that we can expect nothing from them, and it was striking to me a few days ago to hear the "German Daily Paper" and "Forwards" mentioned in the same sentence. They really fit so well together! It is not with a light heart that we pass over the posture of the German people in decisive questions of their fate. We see again with frightful clarity today that throughout the years of national massacres, throughout all the humiliation, lies, and violence that the German people found acceptable, nothing has been altered in the thinking of the people or in its attitude to the principles of justice and morality. The spirit of 1914 has begun to triumph over the revolution! Regardless of their political position, sincere and honest people who set love of truth above all else will always find common ground to make possible the judgment of events, conditions, and necessary tasks.

Whoever clearly sees today the combination of circumstances in which the development of the political struggle in Germany stands will—and I repeat, regardless how he thinks politically or to which party he may belong—be astonished and disgusted by the discovery that all the political parties, even though in many respects they formally contradict one another, pursue one and the same mode of joining battle: by the denial of any sort of morality and by a disposition towards the most extreme materialistic basic principles, the meanspiritedness of their conception of human responsibility, and to some degree a perhaps unconscious denial of all forms of Christianity.

For many years murder and hatred was preached to the German people, and one could hardly imagine a more obedient student! Only when the scales of fate tilted in favor of a coalition of nations that was, fortunately, more powerful—nations which, so far as their constitutions made this possible, were peaceloving and inclined to save themselves and mankind from bestial rape—only then came the swift awakening, and revolution was christened. Surely no one, least of all a revolutionary, could ever have imagined that revolution would take this course within a people so justly defeated. What we are today calling revolution is in fact nothing but a temporary sobering-up, a recognition that for the time being we cannot continue along this path of accomodating to the new situation with all our means.

Thus far, I am convinced, any objective thinker will agree with me that nobody can claim to observe constructive outcomes, or be in a position to show any demonstrably destructive results of the revolution. So one should not be surprised that bestial crimes are committed before our very eyes with self-confident assurance. I say again—no matter how one thinks politically—our human feeling must rise up in outrage at the events in Munich. Leviné condemned to death, and the sentence immediately executed! That one could expect nothing else from this court was clear in advance—but that the court would nonetheless go so far to block any possible intervention from any direction against the sentence by immediate execution, that we had not expected. The Munich Superior Court knew why it was in such a hurry! Even the paragraph 14 Court in Vienna granted Friedrich Adler "the legal opportunity to move for dismissal, which motion must be made within three days." And that was during wartime—under completely different circumstances! One can approve of Leviné's actions or not, that is entirely a matter of indifference—but the incomprehensible baseness of the sentence lies in the fact that an absolutely illegal court murdered a man without right of appeal despite the virtually unanimous favorable testimony of witnesses. Can mankind really allow this to happen, can it trot along its everyday rounds as though nothing has happened?

We cannot expect of Noske's court officials that they should think like Kurt Eisner, who spoke in Bern about criminals a thousand times more stained with blood: "We do not want at all to stain ourselves by judging these sinners. We are ourselves too proud to be their judges."

But men must, simply must finally become aware that matters cannot continue in this way, that the increasingly crude sequence of such crimes has not come to an end!

A new misfortune already threatens! Ernst Toller has been arrested in Munich. If he, too, is treated in the same way as Leviné, then we know in advance that we cannot deceive ourselves in any fashion about the outcome of the trial. There can be no doubt about the judgment of a Superior Court on Toller! The conviction of a man so thoroughly pure of character, animated only by the highest ideals, would again constitute the weightiest accusation against the whole German people, the weightiest accusation, I say, against each individual who had not done his utmost to assure Toller the opportunity to account for his actions by process of law. I charge everyone, and I hope that my appeal to human feelings will not echo ineffectually, to contribute his share to the elimination of militaristic class justice! Sympathy alone will achieve nothing, action is everything!

Walter Landauer

## 6. WE DEMAND:
## ELIMINATION OF MILITARY JUSTICE IN MUNICH!

*Sozialistische Republik* Karlsruhe I (35) 6/21/19

If the war, despite all its frightful sacrifices, its debasement and degradation of man and mankind should, along with its horrors and the unleashing of all that is feral in man, nonetheless still gives us something, then surely it can at least be the recognition of paths that we may not take, if we wish to fulfill our obligations to our fellow-men, and so also to ourselves. Is there anyone today who does not perceive the curse of force and violence, the curse of hatred and injustice!

If there is one lesson above all that we ought to retrieve from all the sufferings forced upon humanity, which if each and every one of us does not patiently fight against, it must be this, that we constantly strive to acknowledge and to fulfill the most profound mission of man.

What the thundering cacophony of contemporary world events constantly reiterates must be deeply felt and understood: renounce violence and the quest for power for power's sake! Fight against injustice wherever you encounter it, fight for the right and for justice, so that these may be achieved!

But it is our duty to fight here, where we are. We do not have the right to expect humanity from others as long as beasts hold dominion over ourselves. Our friends in all the world know that we are counting on them never for a moment to lose sight of the great undertaking that it is our mission to fulfil. We would like to count on their being at their posts and that they will devote all their talents to drawing us closer to the common goal. But, brothers, the sacred duty falls to us to assure that barbarism disappears in our own house.

Far beyond the borders of our Fatherland all decent men are filled with revulsion at the events in Munich. No justification can possibly be found for the fact that men who have dedicated their entire lives in the service of working people, who have lived only to strive for the improvement of human society, should have been killed in bestial fashion, legally murdered.

From the military, whose essence it is to be parasites on humanity, we cannot expect humane feeling or a sense of justice. The guilt for all these appalling deeds lies with those who still support the military, with those who believe they are necessary for their own security, and with those to whom the actions of the military are welcome. The military party sees in the dawning of a new age their own end approaching; they know clearly that their existence is too much at stake for them not to be on their guard in the most desperate fashion. In our vanguard the military must see their worst enemies, and that they want to eliminate them is in the last analysis understandable. But that there are people who should be called upon to support right and freedom who instead support the militaristic lust for murder, we cannot and will not tolerate. We know that the military authorities who hold Munich's fate in their hands are disposed to continue their bloody handicraft.

Ernst Toller, one of our best people, has been arrested and, like Leviné, has been brought before a military court for sentence. Toller's strongest efforts have been directed to the abandonment of all forms of violence and to working with calm deliberation for the realization of our ideals. It is therefore the highest duty for us all to intercede for Toller and his friends, to demand justice for them. We know what to expect from a military court, and therefore regard it as pointless and laughable to appeal to the sense of justice of its judges. What we demand, however, of the Bavarian government is that it immediately take steps to eliminate the military courts; we demand of the Bavarian government that it immediately declare what steps it has taken to this end; and we urgently warn it to omit no effort that will bring about as rapidly as possible the end of the criminal activities of the military in Munich.

We remind the Hoffman government of the countless voices that have been raised for Toller and his friends internationally, abroad, in all parts of Germany and Austria—in addition to those of the revolutionary *proletariat* throughout the world, in addition to those of all well-disposed people.

Never before have the words of Mephistopheles:

> "Of the right that is innate within us,
> of that, alas!, we never speak"

had greater validity than under the conditions now prevailing in Munich. For that reason we call upon all men who are striving for a new era, for a loftier humanity to support our demands in every way.

Walter Landauer

## 7. THE TRUE PICTURE OF THE GERMANY OF WILSON'S 14 POINTS

*Die freie Zeitung* (Bern) III (50) 6/25/19
By Walter Landauer, Heidelberg

The revolutionary movement in Germany is approaching the end of a period of development that has not taken any forward step, that has brought only misery and ruin. All the present rulers of the German free states are so unbelievably sterile that they would have come to grief on their own ill will, would already have long ago succumbed to it, if they were not mere puppets propped up only as long as they were needed, so as then to collapse swiftly and pitilessly into their proper "drab mediocrity". One would not wish for their continuing for a moment, even though we can see nothing better coming after them, perhaps only much worse. With unchallengeable certainty, with incredibly insolent behavior, the *counterrevolution* is rising in Germany and is arming itself for a powerful assault.

This nation of children and fools, of political seducers and the seduced—to the extent that it favored the revolution—expected miracles of it, saw, with Marx, "the question of property as the fundamental question for the movement", and now finds its hopes disappointed. One cannot emphatically enough stress how large is the share of responsibility for the systematic suppression of a revolution of the spirit that must be borne expressly by the *socialist parties*, capable only of worshipping the slogans of their party idols.

The impotence becomes more glaring from day to day for German intellectual life to strike out on its own pathways that would function to animate and fertilize the revolution. The words of Hoelderlin to his brother in 1799 about the Germans seem by now to have been fulfilled: "They are rooted to the soil, and most of them are one way or another, literally or metaphorically, bound to their patch of earth, and if this continues they must in the end be dragged to their death by their precious (moral and physical) acquisitions and inheritances like that goodhearted Dutch painter ... And as the German feels comfortable only in his own home, even if he is living in an open field, without common sense and an open outlook upon the world no individual life of his own can subsist, and truly it seems that both of these have disappeared in the Germans."

The phenomena of decay, of collapse, and *spiritual* bankruptcy are spreading unobstructedly and are finding credence among the people who clutch indiscriminately at anything that promises rescue and salvation.

A "World Union of Intellectual Workers" with its German headquarters in Munich has adopted the motto: "Let good deeds be the action of our soul," and begins its proclamation with the words: "Germany is the conscience of the world, and the world should not tolerate that German diligence should be the slave of England." And they believe that the intellectual workers of all countries will rally beneath this banner.

With magniloquent pronouncements virtually reminiscent of Ullstein-Scherl, Dr Rudolf Steiner promotes his *Tripartition of the Social Organism*, in which he sees the salvation and solution of all social problems. What cannot be expressed by the most profoundly felt human words we find here miraculously revealed: and we find intellectuals and *proletariat* massing around a leader from whom we have heard only empty words, and who has not shown us that he is qualified in the slightest degree to be the propounder of a pure socialism. (Or should the confirmation of this be the co-signature of one of the biggest industrialists in Stuttgart?)

I have mentioned here two of the typical cases only to demonstrate Germany's truly pathetic impoverishment of mind. These cases could be repeated indefinitely; partly of people who are convinced of their calling, but for the greater part they are charlatans who know that the times favor them.

The German nation seems incapable of liberating itself from a slavish attachment that has become a necessity of life. Just as Germans earlier saw in the Reichstag the incarnation of all their highest powers and capacities, now they worship everything that their National Assembly says, this unimproved child of incorrigible parents.

No wonder, then, under these conditions, that the *German military* believes the time is ripe to assume the *monopoly of power* that was only half removed from it by the revolution. There are far too many indications for any doubt to remain about which plans are to be put into action. One of the most telling arguments pointing in this direction is surely the questionnaire of May 21 circulated by the Army General Staff concerning *Polling the Population about a Possible Resumption of the War*. From all sides we hear calls and proclamations designed to whip up the people to a general "uprising." Of particular note, because of its compelling logic, is a passage in a pamphlet directed against independents and communists distributed in huge quantities by the "Iron Squadrons" (naturally though at public expense and with the approval of the National Assembly?!) , which asserts that "every people deserves the destiny that it has." Why then do these gentlemen then not *understand* that and why we were *rightly defeated* and forced to our knees?

The German people will find nothing in any of these things. The rallying cry of the Volunteer Corps: "Patriotic Germans join Luetzow's wild and daring chase" seems to have recaptured them once again!

That we are approaching the moment—and many signs point to the fact that the deadline for signing the peace treaty will be used as the occasion for this—when once again the whole nation will be called to arms, is illuminated especially brightly by the following anecdote:

A certain Dr. Kolshorn is at the present time making propaganda trips in Baden, visiting the middle-schools(!) and the universities, and giving talks about the measures to be taken when enemy troops march in following our rejection of peace, about the arrangements to be made for those pupils and students fit for service to be enrolled in volunteer brigades; and in the course of this fulminates in the most unrestrained manner against independents and communists. This gentleman travels on behalf of the *Akademisches Heimatbund*, and with *a credential issued by Noske*. Although Noske (just writing down his shameful title is impossible) now denies, in answer to an inquiry by the Baden Minister of Culture, that he has ever issued such a credential to anyone, there can hardly be any doubt that this gentleman really is an emissary of Noske's.

From all this, however, it is clearly apparent that Noske, doubtless in collaboration with the old military potentates Hindenburg, etc., is surely counting on the rejection of the *peace treaty*, without even waiting for the results of the negotiations; and that he is counting, as an excuse for a new recourse to arms, on bellicose actions from outside and, above all, from within the country.

If the German people do not notice, even at this very last moment, in what an outrageous manner it is being driven into new bloodshed, then nothing can save us any longer from a new frightful sacrifice of human life. I do not want here to outline a perspective on all that follows from this, of the civil war that it must provoke, and from which, regardless who wins, no good can possibly come. The most pressing injunction to the German people, and if they do not listen, to the whole world, must be *to do away* with militarism, *to get rid of the military* altogether *as fast as possible*, to deprive it of its means of existence. If the German people still do not see that force can ever bring about any constructive result, whether from the right or the left—and just now more than anything else force from the right is the threat—if they do not understand that we must above all learn to think differently, to respect the humanity of everyone, then its decline surely cannot be halted, *then, perhaps, it is not even desirable to forestall this decline*. We shall soon see if, with F.C. Endres, we may still hope:

May spirit live, it leads to highest goals,
A painful path through sodden earthborn night,
Go forth with joy! The mountains are alight
With brilliant sunshine, and the hope of souls.

Walter Landauer

## 8. IS THE QUESTION OF WAR GUILT SETTLED?

*Die freie Zeitung* (Bern) III (58) 7/23/19

If we feel it to be our duty to point repeatedly to the guilt of the German people, to the exclusive guilt of Germany for the outbreak of the war, and to the overwhelming guilt for the decades-long stimulation of bellicose thought, it is because, and this must also be emphatically stressed, it has wide-ranging significance. From all sides in Germany one hears voices raised against one that it is irrational, insane to continue to talk of war guilt at all, now that the "degrading" peace has been signed. If we look more closely, what we find of course is that it is just those who would reproach us and represent it as a weakness who are responsible for characterizing us during the war as criminals against the German nation; they are the ones who find it a paradox, or pretend to, if anyone speaks of German guilt at all. All of us who knew, from the first day of the war, and even before, what would be involved in a German war and where it would lead, were treated as "belittlers," as the spiritual source of Germany's military defeat, as the harbingers of revolution.

We cannot be saddened by this insult which does not alter our convictions by an atom. What is necessary today once again is to establish what the issue is when we renew over and over what is at stake when we speak of German guilt. In the earliest days of the revolution we too abandoned ourselves to the hope that it would no longer be necessary to talk of war guilt, that the German people would have come to its senses, would have woken from its benighted condition, shuddering at its deeds, and disavowing all its war criminals and warmongers, and so strike out on a fresh path. *Even the least of our hopes has been disappointed. Our revolution comes first, must come,* if mankind is to emerge from the abyss into which it has all but fallen.

Why, during the war, we fought against militarism and the war psychosis, against the passion for expansionism and pan-Germanism, need not be repeated now. One word should suffice: *Truth.* And, indeed, it should still suffice today. Those who are ready to condemn people who think differently, even before they have heard what they say, are lost to the cause of mankind. Only those who have probed to the sources of events, those who have taken serious and implacable pains to overcome the prejudices and distortions in their surroundings, are capable of grasping the inner coherence of historical developments. This is the way we see the situation today:

Under the spell of ideas forced through education and coercion for decades the German people—or a decisive majority of them at least—has been unable to liberate its spirit or itself. The peace treaty was signed, not as we hoped, not as was self-evidently to be expected, as the termination of a superseded era, as a reckoning with all the criminals who had plunged Germany's people and mankind at large into nameless misery for years, as the foundation stone for a generation of spiritually renewed men, and the sincere collaboration of peoples for which we strive. Peace was instead signed by Germany in an outrageously frivolous spirit poisoned by prewar thinking, which sees in the treaty nothing but the blackmail imposed by force by the "enemy."

Once again there opens up the deep, seemingly all but unbridgeable cleft that separates German thought from mankind. Just as in the final days before the war, the German spirit reveals itself once again as overbearing and arrogant, rejecting the best efforts at compromise, while its opponents are motivated by a desire for reconciliation and love, taking the comity of nations as its target. Anyone who did not sense the profound moral gravity of *Lloyd George's Note* of June 16 will hardly be able ever to grasp it, so far has Germany's thinking fallen below that of other peoples. This Note is elevated well above Wilson's addresses to Congress. Perhaps never before in the history of mankind, certainly never in time of war, have such words of sincere, solemn desire for reconciliation been addressed from enemy to enemy. Incompletely and badly translated,[1] this classic document was placed before the German public, which for the most part is unaware of it, or knows of it only in the commentaries of the German press, amply renowned for its immorality.

"We yearn for the rapid extinction of the passions inflamed by the war, and for the community of nations without exception enjoying the prosperity which the honest satisfaction of needs will assure, wherever they may arise. We do not begrudge the full enjoyment of this prosperity to the German people, as to every other." Every word of the Note speaks to its intrinsic truthfulness. In it is rooted a demand laid upon us that we cannot refuse, that must lay the foundation for all our future existence, for any further development and acknowledgement of its own guilt by the German people. "Somebody has to suffer from the effects of the war. Who should it be? Germany, or only the peoples for whose misery it is responsible? If, in the near future, there is to be talk of reconciliation and mitigation, this will be possible only if those responsible for the conclusion of the peace can summon the courage to

---

[1] This was the topic of several contemporary stories in *The New York Times*; the German and Austrian Foreign Offices are reported as responding that they had no obligation to lay the enemy's proposals before their peoples. The citations from *Lloyd George's Note of June 16* appear as a lengthy summary in *The Times* of London of June 17 1919—it was signed by Clémenceau—and refers to the full text of some 20,000 words. Of the passages cited by Landauer, the only one which appears to be a direct quotation comes from the summary. The rest no doubt comes from the full text which we have not consulted. The passage below: "Somebody has to suffer ...it is responsible" has been translated back from Landauer's version. The actual text (*Times* 6/17/19, p. 17, col. 2) reads: "Somebody must suffer for the consequences of the war. Is it to be Germany or the peoples she has wronged?"

assure justice in all its stark purity, and not sacrifice it in favor of a comfortable peace settlement."

Who were those responsible? Was it only the German peace delegation, only the National Assembly? No, *the responsibility today is that of the entire German people* that found itself unable to find worthy representatives with clean hands who could justifiably make its case. Nothing, absolutely nothing, was done for the cause of mankind when the peace treaty was signed under the relentless coercion of military impotence, and if its provisions were to be implemented in the same way. Just as during the war, and during the cease-fire, it is even more the case now as a foundation for its new existence and future that Germany satisfy the demand that seems to me the most important.

The final words of Lloyd George say: "And the treaty must be accepted or rejected, both in letter and in spirit, just as it stands." That has not happened; that is what we still strive for. Peace has been signed according to the letter, but acceptance of its spirit is lacking, as is any mental concurrence at all. This spiritual foundation for the new age at the beginning of which we stand cannot be introduced into the German people from outside; it must grow from within and determine all its actions. So far the revolution has not been able to fulfill this, its most important demand, and for this reason we shall incessantly place in the foreground the complete transformation of the German mentality as the most important demand. One part of this demand is the acknowledgement of German war guilt, and of the consequences that flow out of it, for, as Lloyd George says: "The essence of justice demands expiation for injustice done".

*Walter Landauer*, Heidelberg.

## 9. GERMAN PROFESSORS IN THE GLARE OF POLITICS

*Die freie Zeitung* (Bern) III (59) 7/30/19.

Much discussed, much denounced—it may perhaps seem unjust once again to pick out the political views of German professors from the circles of politically active and politically inactive people, and to use them as a terrifying example. And yet it seems to me, there is no better mirror for the German people to see itself and its products than precisely in its intellectual aristocracy. All malice to one side—I know that there are exceptions and will stress this—but the very prototype of the politically extreme, onesided thinking German is manifested in the academic teachers of Germany. Here we find in the most emphatic form all those characteristics combined that have isolated Germany among the nations, that are driving it closer to decline, inspiring the decaying structure of a mode of thought already long condemned to collapse.

As the revolution stands today it can hardly be questioned that in the foreseeable future a truly revolutionary spirit, which we have long awaited, will arise among broad circles of educated people in Germany. The revolution advances without doubt—and whatever it will bring us, whether for good or ill, it cannot go backwards. Today the responsibility for our future rests in large measure with the revolutionary *proletariat* at whose side only a few members of the "educated" classes stand in this grave battle. *But it is necessary to increase their numbers!*

If we return once again to point out the peculiar role played by our German professors in political affairs, we do so for the purpose of waking these people up. To believe this possible can apply only to very few of them. But it is our principal task to help the many who half unconsciously seek new paths in their search. I am well aware that the effort that we undertake here will produce little of direct positive value. But how little of positive value are we able to achieve today altogether! Our activity, if it is to have some success at least, must now limit itself in many respects to demonstrating how all approaches tried to date are—fortunately—hopelessly destroyed. Above all, what is needed is to secure the belief, and gain ever more adherents to it, that we must strike out in wholly new directions. Once we have gathered enough people around us who do not want to continue to live this way, and who are gladly prepared and determined to stand with us in the endlessly difficult task of reconstructing the history of humanity, then we shall be delivered from the worst.

If one attempts to characterize the university professors of Germany in accordance with general political perspectives, then it seems to me that we always find four points of view, four different sharply limned types, who are in one degree or another in opposition to each other.

First among them are those few who, throughout all the tempests of pan-German power politics and systematic lying and vandalism, have preserved, despite the most oppressive personal sacrifices, a love of unconditional unqualified truth, and the courage to oppose crime and violence. Professor Nicolai, Professor Foerster and a few, very few others, are those I have in mind here.

In terms of professionalism, which it seems to me in large measure belongs to the historical period we have left behind us, it is still plausible to include the fairly large number of professors who at least attempted—they could only rarely succeed under the existing circumstances—who honestly tried to remain neutral. Even if under some conditions inactivity is worse than fighting against the good, still it must be acknowledged that many people, and they were not the worst who, conscious of their own political inadequacy, could not make up their minds. Among them we must also include those who made it their duty to distance politics from their teaching, who were aware of the enormous responsibility that in whole or part burdens those who, under cover of science and public opinion, served party politics or racial prejudice. Even though we cannot honor and love these people as fellow warriors in the cause of humanity, we cannot deny a great many of them our respect at least. But we must join open battle with those we call pan-Germans and ultra-pan-Germans, with those who even today find it necessary to rape mankind with the barbarity of their political views. How this pan-Germanism that I am referring to expresses itself we saw best in Houston Stewart Chamberlain. Nothing would be easier than to cite innumerable examples of the same from the German professoriate, such as Professor Lenard, alias Lenardi, one of the signatories of the Declaration of 93, who to this day enjoin their followers "to reject far from us all forms of internationalist thinking." How far, however, this kind of "politics" is prepared to trespass we would find it hard to believe if we were not daily obliged to experience it. It sounds like a fairy story to us to hear that a German university teacher distributes to his seminar (on "animal genetics") a political broadside by the tenured professor Kerschensteiner (!), that he recently described Kurt Eisner as an arch-scoundrel, and asserted that the happiest recollections of Germans are associated with their successes in battle. Despite an admonition from the appropriate Ministry of Culture this gentleman does not find it necessary to temper his activities. He grew up in Switzerland, taught for a long time in Swiss universities, and during the war asserted in his lectures on "Attitudes in Switzerland" that the great majority of the Swiss people thought "as Germans"—politically, of course.

Professor Hermann Kraemer, of Hohenheim, have you never found even a glimmer of understanding of the profound truth of the words of Theophrastus Paracelsus of Hohenheim: "The arts are not sealed up in some Fatherland, but are distributed throughout the whole world; they are not in some individual or in one place, but must be gathered together and sought out, wherever they are." And again: "A text is explored letter by letter, but nature from land to land, and for each land is a separate leaf. This is the Codex of Nature, and so must we turn her leaves."?

Faced with such characters one is involuntarily reminded of the fine saying of Richard Dehmel from his prewar days as a delivery man, that he must thank ten nations for the little brain that he has. Evidently, we can find no other explanation than that these gentry felt obliged to repay their various loans no later than the 1st of August 1914. And what are they left with? It makes one shudder!

We cannot surrender to the hope of bringing these gentlemen to their senses but we do feel the obligation of drawing attention to their mischievous activities, to the curse that lies on their work. One of the first tasks of Germany, one of the most important duties of our coming revolution, must be to find pure educators who possess science and art, and for that very reason religion also, and for that very reason also deepest love for all mankind.

Walter Landauer, Heidelberg

## 10. COLLEGE REFORM

*Die freie Zeitung* (Bern) III (67) August 27 1919
By Walter Landauer, Heidelberg

Not a week passes without the appearance of a quantity of tracts and books dealing with the problem of reform of the German colleges. In many circles people seem to have the feeling at least that few things in Germany are as run down as the public institutions of higher learning. No wonder then, that an intense battle has been joined about the renewal of academic stools of learning, as about everything else; but one which fades into pedantic meaninglessness. Yet how ludicrous it is for thousands to engage themselves in renewal when they should see at first glance that renewal is altogether impossible. Questions of preparatory education, the examination system, of college fees and the manifold other questions cannot be approached from the perspective of improving existing abuses; this produces only *cuckoo's eggs like the new Education Act*. Those who are seriously committed to the emergence of a completely new education that will lay a foundation for our entire intellectual life must be told in the most emphatic terms that under present conditions nothing can be achieved, that even worthless patchwork is barely possible. As long as we are as far removed from true democracy as ever, any attempt at transformation of a partial area of human life must remain hopeless. It is in any case clear, given the intimate interrelationship of all the conditions of life, that a piecemeal approach can hardly bring success. All the less, because *the revolution has all but expired because of the spiritual lethargy of the German people*, and we do not know when the revolutionary spirit will flare up anew and bring about great events, a completely altered mode of thought; and it is for this reason that we feel called upon to prepare the way to herald the time that will realize the fulfilment of our ideals. It is certainly important to reiterate over and over all the problems that the revolution has revealed afresh, and to the solution of which not even an attempt has been made, if only for the purpose of bringing them to the consciousness of the public. For all that so much is now spoken in Germany about *college reform* on the one hand, and about *comprehensive schools* and other pedagogical problems on the other hand, how little attention is paid to the indisputable connection that obtains even among these problems.

Even if one believes, as I do, not only in the slow and steady evolution of human history, but also sees at work, alongside that, discontinuous leaps in the advance of civilization, even if only rarely, one is still obliged to view the *resolution of problems of education*, the most widespread prospect of creating a new and purified mankind *as one of the central problems of our time*. In fact, there is only one way out of the labyrinth of questions how the new education should be actualized: the German people must first undergo a complete change in their outlook; only then can goodness take root, and education become possible and fruitful.

## 11. AGAINST KARL MARX

*Die freie Zeitung* III (74) 9/20/19

A book has recently appeared which, more than has previously been the case, warrants that it should be widely noted and read. I am referring to the work by Pierre Ramus (Rudolf Grossmann): *The False Doctrine and Unscientific Nature of Marxism in the Realm of Socialism* (Published by Erkenntnis und Befreiung, Klosterneuburg/Wien). This book by the editor of the newspaper *Erkenntnis und Befreiung* deserves the widest notice for a number of reasons. Above all, it seems to me, that the book advances one of our most important tasks and fights against one of our greatest evils. The damage that Marx has done to workers is of immeasurable dimensions. How completely differently, without a doubt, would the workers have taken a stand in face of the threatening danger of war, if their thinking had not been filled with a rigid, lifeless and mechanical dogmatism, if they had not made into their most precious possession a bible overflowing with heaven-storming, sacred idealism that acknowledged no restraints. The world will not be shattered and put together again by a belief in the unfolding of history in accordance with a law—it will only be so through an unqualified belief in man, in his knowledge and will, in his calling to goodness and wisdom supported by rationality.

These are the ultimate dichotomies that set the existential perspectives of Pierre Ramus worlds apart from those of Marx. It is not just a question of rebutting individual points of Marxist doctrine; rather it is by *demonstrating all its elementary errors* to cut the ground from under it in order to make room for more fruitful ideals. Whoever knows that true socialism is "ahistorical and antitheological," that it can have nothing to do with Hegel and his quintessence of philosophy: "the real is rational," has long since clearly understood that our three socialist parties display only differences of degree, but are in essence Marxist, i.e., unsocialistic, not aiming at the liberation of all men but only at the overthrow of social relations and restoring them only in favor of the few. Marxism and the Marxist parties are not engaged in overcoming *Campanella's* deeply felt insight: *"All evils spring from the opposition of wealth and poverty"*; their aim is not to liberate people by and with people, all people—but rather to establish a "revolutionary-proletarian-absolutist State power," an "institution lying outside working people themselves."

Of his work, Ramus rightly says that it is the first to concern itself in a readily accessible manner with Marxist criticism in the spirit of socialism. From a bourgeois point of view Marx has been subjected to many criticisms, some brilliant ones among them, led by Franz Oppenheimer. Far fewer opponents have stood against Marx from the socialist camp, for nearly all of them submitted to his fantasies; when he was confronted it was mostly only to make room for new ideas. I am thinking here above all of the *Call to Socialism* by Gustav Landauer. Ramus devotes himself only to the criticism of Marxism, to the shattering of the desert of Marxist ideas, and he is particularly to be thanked that he has done this in a manner that is clear, objective, and understandable by everybody. The parties, of course, are dazzling in their ability to keep anything away from the workers that might serve to open their eyes, and for that reason we hope that everyone who agrees with Ramus—if only in their opposition to Marx—will do all that is possible to spread the word about the book and its content.

Ramus' criticism is limited in the main to the *Communist Manifesto* and to Engels'*Anti-Dühring*, and displays in a most striking manner the innumerable contradictions in which Marx and his great pupil entangle themselves. His *criticism of the materialistic philosophy of history* is truly annihilating, and his demonstration is irrefutable that Marx is no more a communist than a socialist, rather—above all in *Kapital*—that he is a theoretician of national economics.

From each sentence of this book we gain the sense that here we are addressed by a sincere socialist, and anyone who is close to Ramus' aims will joyfully hail the indications he gives here and there of his own constructive thoughts, instinct with the spirit of Proudhon and of Tolstoy.

The final words of Ramus' book are the best testimony of the sentiment in which this work was written:

"*Only after the complete defeat of Marxism will socialism awake to new life.*"

Walter Landauer, Heidelberg

## 12. THE PUBLIC PROSECUTOR

*Die freie Zeitung* (Bern) III (78) 10/4/19

Conditions in the German Republic, the treatment of political "criminals," cause many people who view these occurrences critically to meditate about the rights and duties of the "District Attorney," the public prosecutor, the upholder and defender of peace and order. Since all *a priori* generalizations about classes of individuals are unjust, one cannot simply attack all the public prosecutors in the German Republic without qualification. It is implicit in the nature of a judge's office for the judge to be exposed to the danger of deciding cases, not in accordance with individual circumstances, but in accordance with the law, and *only* in accordance with the law. Strong personalities will always find the right modality even here; but that there are only a few of these is the fault today less of the individual than of their training.

If the reputation of the public prosecutor was not high even in earlier times—among the people as little as among jurists—the revolution and the jurisprudential processes connected with it have completely undermined public confidence in the office. To be sure, the present *régime* makes use of its power, aims at surviving, and will not condemn itself. To be sure, the erstwhile faithful servants of imperial justice will not suddenly become advocates for the highest humanity. But none of the preconditions have been met that would display even the intention to raise jurisprudence to a higher level. The death penalty does not apply only to vile crimes. Unwelcome beliefs, contrary political convictions suffice to extinguish human life.

The military court has disappeared in Bavaria, to be replaced by the "People's Court," which differs from it in no substantial way. The everyday business formalities of agriculture, trade and industry are conducted through innumerable levels of jurisdiction, demanding an army of scribes and officials; but in trials affecting life and death in which the danger of subjectivity should be avoided more than anywhere else, power rests in a single hand, and the accused is deprived of the possibility of appeal. Asserting that, for example, the sentence of the Bavarian "People's Court" must be confirmed by the Ministry, has unfortunately no serious significance.

There can be no objection to the fact that the murder of the "hostages" in Munich has been painstakingly investigated, and the perpetrators brought to justice; but one can have a very different opinion about the conduct of the process. Alexander Herzen's dictum applies to it: "For them the judicial process is only a cruel game, like bullfighting for the Spaniard. The public prosecutor feels as humiliated and insulted as a skillful toreador, if the tormented animal is alive at the end." The prosecutor did not make the least effort to challenge the guilt of the other party. Just as in the Ledebour case the prosecutor felt it necessary to take his client, Count Westarp, under his personal protection, describing him as as an individual of indubitably elevated morality, so too in the hostage murder case it was taken as an *a priori* proposition by the prosecutor that the murder victims were angels. Throughout the trial no light was cast on the anti-revolutionary secret conspiracy of the Thule Society, the counterfeit authorizations were swept into the background, and these as well as many other elements excite the suspicion that it was a case of political prisoners rather than hostages.

Reason has become nonsense, beneficence a curse;
Woe unto you, that you are a mere descendant![1]

But the Bavarian government has so far done nothing to investigate or clear up all the innumerable crimes as a result of which many of our best people have lost their lives, including many who were not active politically at all.

We are still far removed from Tolstoy's moral dictum: "See that you so live that you have no need of violence." Yet how easily could we move a sound step nearer to him if people could only gain insight into the words of Thoreau: "The law has never made men more just; on the contrary, good men fall prey to injustice because of respect for the law." Nothing is more apt than the terror of these days to cure people of their exaggerated respect for Draconic law and for a brutal executive, and to lead them to establish a higher forum of justice before which they would judge human actions by another yardstick. The Ledebour trial demonstrated with especial clarity how extraordinarily important it is for laymen and professional jurists to work together, to complement each other. Only in this manner can the public prosecutor suceed in judging his victims from a more human perspective. Then, too, can Kant's adage about jurisprudence in its battle against evil be given reality: "Virtue is a moral attitude struggling with itself."

Walter Landauer, Heidelberg.

[1] Goethe: *Faust I,* lines 1976-7. In Charles Passage's translation:

Sense turns to nonsense, wise works to a mire.
Woe that you are a grandson and born late!

The quotation is apposite because Mephistopheles is telling the Student that laws and law schools decline from generation to generation, so the later you are born the worse off you are.

## 13. DESPOTISM WITHOUT END!

*Erkenntnis und Befreiung* (Vienna) (I) (45) 10/15-21/19

For Wilhelm Herzog and the "Republic"

"Lest you see it too late, alas!
The watch on the Rhine will not suffice,
the worst enemy stands on the Spree ..."

Herwegh, not a prophet, but the accuser of his age, is as right today as he was then. The causes of the revolution of 1848, the law against the socialists, the German republic of 1918—represent a single line of development. Friedrich Wilhelm IV, Wilhelm II, Noske, its representatives! The leaders of the present government in Germany suffer the curse of today and our generation the contempt of the future. The Social Democrats, the opposition of the previous government, gave the German people a new constitution. Germany is now—as is well known everywhere—the "freest democracy in the world." But how this democracy operates can here be illustrated by a single case.

Revolutionary workers in Berlin, forming themselves into a *cooperative*, created in the *Republic* an instrument of their struggle that was widely circulated as a *"newspaper of the German workers' councils."* Not only did it not serve any party's aims, the *Republic* distinguished itself from virtually all opposition newspapers by not adopting the provocative and street jargon of the party papers; that, it seems to me, was the best thing about it. The *Republic* must have become accustomed to temporary bans, but on the 23rd of June 1919, it was banned *"for an indeterminate period."* Noske—the "supreme commander in the shires"—ordered it.

As was shown in the case of Reinhardt-Neuendorff, Noske from time to time evidently believes it necessary to conceal his despotism from the general public. This was so in the banning of the *Republic*. In the National Assembly Noske sought to justify himself by *deliberate slander* and underhanded abuse of his opponents. He characterized Wilhelm Herzog and Ernst Däumig, the editors of the *Republic* as people who "have only just discovered their revolutionary hearts, and now want to cut political deals." Indeed—what bitter irony—he characterizes those people who boldly stand up for their convictions, regardless of when and whence they have come, if only they are sincere, as "*carrion vultures of the revolution.*" *Condottiere*[1] Noske is truly in a position to speak.

That the editors and collaborators of the *Republic* are outraged at such blackguardry should be understandable by anybody. And yet how little has

[1] Italian: leader of a band of mercenaries.

happened so far for someone to come forward in favor of right, justice, and freedom. The absolutist *régime*, which dominates the people arbitrarily and unconstitutionally, believes Germans to be naive enough to accept this—but much worse than that is a *régime* that pays no heed today to the laws that it passed itself yesterday! The constitution of the German Republic does not permit "censorship"; despite this, the ban on the *Republic* has so far not been lifted. And the natural consequence of this: the people "who want to cut political deals" are financially ruined! The ban on the *Republic*, that has now lasted over a quarter of a year, has had, as the unavoidable result, that it has collapsed. Thus we are not surprised to hear that bankruptcy has been declared on the assets of the publication cooperative.

Whether Noske counted on this in advance, or whether he now rejoices in the results of the ban, is neither here nor there—but we hope that the solidarity of the readership of the *Republic* will cause it to reappear shortly.

One should not believe that the strangulation of the *Republic* is wholly unknown to the German nation. The entire German press carried the news of it, some merely recording the story, others displaying malicious satisfaction. And beyond this, Wilhelm Herzog defended himself against Noske's attack in a letter to Fehrenbach, seeking his protection. Of course, Herzog was bitterly disappointed when he believed that he could count on the "sense of justice" of the president of the German National Assembly (a lawyer and at-large representative) to *"protect him, a completely defenseless person, from outrageous insults."* On the contrary, Herr Fehrenbach surely regards it as entirely in order "that a minister can utter slanders without fear of contradiction." The National Assembly falls silent, and then talks on—so far revealing as little of its democratic as of its Christian heart. Noske's puppets will continue to dig the grave of the German people...

The ban has, however, done the greatest damage to the readership of the *Republic*, because it has forbidden the exchange of ideas. May Noske's power fiat demonstrate at least one thing to them: that no blessings can flow from the exercise of violence, that dictatorship means the suppression of equal rights—then perhaps something can be gained, and Noske may have done good despite his own intentions. They might join Lermontov in saying:

" If I am often embarrassed by your actions
And the jangle of chains does not sound like music,
And I am not lured by the flash of bayonets;
Do not suppose that I am not a patriot!"

And the day will dawn for the *Republic* only when Germany itself *really* becomes a republic. Then we shall be a step forward along our difficult path. May that day soon come!

*Walter Landauer (Heidelberg)*

[#14 Review of Hugo Ball omitted]

## 15. THE FIRST CONVENTION OF THE GERMAN NATIONAL PARTY

*Die freie Zeitung* (Bern) III (85) 10/29/19
By *Walter Landauer*, Heidelberg

On the 12th and 13th of October the *German National Party* met once again in Berlin. We know that the number of its local chapters rose from 2284 to 3024 between July 1 and October 1, also that the number of registered members of this party passed the 1.1 million mark during this summer. The object of these lines is to pick out some of the "high points" of the convention, and to draw some parallels. The entire assembly was highly satisfied with the course of events, and that is not surprising. Although there was little sympathy indeed among Germans generally for these antediluvian people, still some slogans were coined that are to a high degree indicative of the present state of their development.

In his opening speech State Minister Hergt asserted: "No party but ours can pride itself on reaching so deeply into all classes of society." "We are not given to over-hasty actions." No one, indeed, would dispute this. At least no one would hasten to fall forwards in any case—falling backwards is more likely, you fall on a softer part. "We are free of partisanship." What you do know! "At the present time, when the parliamentary system has been forced on us, the only important thing for a party is to achieve a majority. All other objectives recede in importance." Such confessions are always disarming. "We cannot do without a strong, undivided Prussia . . . *so that the qualities of character that can flourish only under Prussian unity, and of which we are so proud, are not lost to the German people.*" Dr. Mathies did not fare so well as a fellow-speaker: "Hearts are soaring toward us all by themselves." Well, that is something we don't believe, in the first place because it is clearly established that in Germany hearts unfortunately have a very hard time flying, else we would have left much evil behind us. "Imagine to yourself a German National man or a German National woman, filled with a true spirit of Christianity, with burning nationalist feeling . . ." Just imagine that, if you please! Even if the entire German people were seriously to tax itself just trying to imagine this, it would not only discover the nullity of the German National Party, but also its inadequacy to finding new paths. "We still do not have a socialist republic," falls somewhat soothingly from the speaker's lips, and he then mints his finest dictum, he speaks of the "unbridled lust for freedom" of the Social Democrats. We are convinced that Herr Noske will appropriate this testimonial for himself as soon as possible. "A people gathers true honor for itself through its character, through its conception of service to the people, of service to mankind." Hear this, you Germans, you can actually learn something for once from the German Nationalists! We can hardly say what we believe any better. Who is there in Germany who is concerned for the welfare of his people, or of mankind—I would almost add, who is concerned for its character? We're still looking . . .!

Representative von Kardorff said, among other things, "If fate would have it, given the unfolding of events, that this Prussia should be divided, then speaking for myself I would prefer a centrally administered state to a Federal one." That is something that many "Socialists" in Germany should think carefully about. Why do the German Nationalists advocate centralization; why must anyone who wants freedom advocate decentralization? Think carefully about it!

Dr. Steiniger: "With all respect, we miss a great deal in Weimar, a great deal of the old Prussian spirit." Here the emphasis is supposed to be on the respect, yet the old Prussian spirit is not so lightly killed off.

Helfferich responds to an interjection that the meeting is not a hanging tribunal. Indeed, but that is taken care of just as well by other people. The old saying *cuius regio eius religio* has been changed into "whoever runs the government possesses the right to hang." This, despite the Erfurt Program, is democracy and Christianity! Helfferich's assertion is interesting in that already in 1915, as Treasury Secretary, he had preparations made for a tax on capital. Did Helfferich reckon back then on the possibility that the game could turn out badly?

Pastor Traub, the *Eiserne-Blätter* man and a noble servant of God, spake thus: "What in the last analysis is the League of Nations? For all practical purposes it entails nothing more than the spine-stiffening of France. For France still fears us to this day." Nobody would deny that that the League of Nations is well on the way to becoming just the opposite of what it should become. And in this these sinister men of honor bear their own fully packed measure of guilt. Further: "they're still afraid even today"—that is the typical phrase of provocation. That is what all the military in Germany believe. And with it they think they can at any time clothe the people in armor anew. The following fits this to perfection: "It is my conviction that *the old honorable Prussian State*, historical shortcomings to one side, *was the most social state in the world.*"

Unfortunately Traub has nothing to add about the "historical shortcomings"—and surely it is just those that make Prussia seem to us the least social state. Yet Traub immediately confesses: "I have no new ideas." Note this once and for all.

Von der Osten accuses socialism of acting in accordance with the saying:

And if to be my brother you reject,
To smash your skull in I elect!

But—"duty will keep our heads above water, our trust *in God and in our great old Prussian spirit.*" Prussian spirit, for God's sake! How few there are in Germany who still think that it was just this Prussian spirit that, alas, killed so many.

Winckler poses it as a "fundamental element of the German National outlook" that "we want to be German Christians," since then we would have "the most powerful authority with us in support of everything that we want to do." With God's help they want to do everything that serves their purposes and leads to the degradation of mankind. That is the only way to understand it. For, if they believed that God can do everything to them, they would not be so excessively loud.

What illustrates this to perfection are the words of Delbrück: "What first unites us is the monarchistic principle, secondly the Christian outlook, but surpassed and dominated by the belief in the necessity of a strong national state." How pathetic these people are to say such things, let alone think them. No one demands that anyone must attain or even become like his ideal. Everything depends on good will, on an earnest striving. But this—to propose an ideal and immediately, in pursuit of one's own interests, to set its antithesis above it—that is the epitome of immoral character. Nowhere more than in religious matters has this inner contradiction in the German people taken root.

Pastor Reinhard believes it is necessary to recall "a man whose spirit is among us, Fichte." Poor Fichte would surely never have suspected this, even though he shares the blame. But enough, more than enough!

We flee these things at first sight.
A garbage bin and a junkyard,
And most of all affairs of State.

Just believe your eyes, all these things just happened. And a great deal more. I have not even allowed Hugenberg, Count Westarp, Maurenbrecher & Co. to be heard from. Whatever for! These quotations suffice to give a picture of the fanaticism that reigns here. It is obvious that much was spoken about the Hohenzollern spirit and policies; that drivel in all the keys of the scale was uttered about Wilhelm the Dethroned. But it is much more important to recognize how great the damage is that proceeds from this. Even if the aims of these individuals can never again be realized, yet the harm is immeasurable which comes from them. This is what needs to be fought against.

## 16. THE PARTY

*Die freie Zeitung* (Bern) III (92) 11/ 22/19
by *Walter Landauer,* Heidelberg

"Centralism is stupidity," Oscar Wilde once said. This insightful dictum casts a bright light on many of the problems of our day. Centralism and stupidity—they are interchangeably cause and effect. They breed incapacity, the failure of every initative. They suppress personality, inhibit the unfolding of the innate values of the individual. They interpose iron barriers in every dimension to the development of mankind, fundamentally eradicating every impulse to a moral life.

Centralism is the denial of the divine in man, it is servitude and slavery. To fight against centralism in every form is to fight for the self-knowledge of the individual as unique. The conquest of centralism is the self-discovery of all those people who have recognized their highest vocation. And this is the source of community. We hardly know how splendidly community will express itself, for we do not know it. But it can come about only by clearing away the rubble of centuries.

Not in the slightest has the party anything to do with community. Community is harmonious, the party is disharmonious. That harmony can arise only through the association of people with similar objectives in their fundamental thoughts and desires. Political life today is dominated down to the last detail by the party. *Everything that is said here therefore applies regardless of their programs to all parties without exception.*

The concept of party appears to our contemporaries to be something absolutely necessary. They can hardly conceive any more of politics without a party; they believe it to be impossible despite the youth of parties in human history. An authentic child of the 19th century, the party has attained its adolescence in our century. Like almost nothing else it is the expression of our time which strives toward compactness and mechanization. We hope that its old age and demise are close, and believe that there are many indications already to be seen of this.

The party is supposed to be what it never can be—representation of the interests of certain classes of individuals and social strata. People of like disposition indeed wanted to associate themselves with it. It is understandable for this reason that the concept of a party sprang from revolutionary circles. And in fact we find its first blossoming in the French revolution.

From what has been said at the outset it follows that there is no objection to the association of creative and active people of like disposition—they will thereby complement, support, and enlarge each other. This, however, was only the initial origin of the party; this was the way individuals found their way to common action. The comradely association of people who no longer wished to live this way—who could have anything to say against it?

The nature of the party has nothing to do with this. Quite apart from the dirty propaganda with which all the parties have recruited their membership, it is in principle impossible under the circumstances of the day for all members to square every action of their party with their consciences. The many schisms in the parties today speak volumes of this situation. This is disharmony, and where it reigns no good can come of it.

But it is just these schisms, this collapse that are the better indication because they are evidence that people are thinking for themselves, even if incorrectly. But by far the greatest number of them who proudly call themselves members of one party or another are still sworn to uphold some sort of program, the platform of the party. It is therefore understandable that they are still guided only by this program, and that in this respect they lose sight *of the purely human interests that stand as heaven to earth above all others.*

Every political party within parliamentarianism must pursue one constant goal: to obtain a majority. How else could they expect their program to be actualized and not merely represented? But what does that mean? What it very plainly means is that *each individual party, whatever its name, aims at dictatorship and oppression.* For surely it is obvious that each party will impose its own thoughts as far as possible, and will ignore the rest. Until the revolution the move in Germany was to the extreme right; in Russia it is now the party that stands farthest left that finds itself in this situation. And it is for this reason that everyone who fights against dictatorship of every kind finds all political parties impossible for them.

The leveling-down of the ideas of Christ was the Church, and the leveling-down of all political ideas is the party. Just as Christ's teachings must become the experience of each individual, if they are to reappear in fresh garments, so too can innermost experience lead to a change in political attitudes in order to bring them into accord with mankind's highest interests.

The essence of the party is the incapacity of the individual; but the experience of each individual becomes the highest strength of the whole. The party is an unmistakable sign of decline. *Every party, as a party, is already conservative.* I see only one path open, the sublimest that mankind has ever trodden, that must become the path of the destructive, restless and undivided experience of the individual: *the party of mankind*.

## 17. THE FREE GERMAN YOUTH

*Die Menschheit* VI (13) 12/27/19
By Walter Landauer, Heidelberg

If somebody were to ask me which were the best and the worst books of the last few years, I would be certain of one thing, and that is that the answer would fall within a narrow range. Without hesitation I would hold Leonhard Frank's *Man is Good* to be among the best—and among the worst Hans Blüher's *In Medias Res*. I do not mean to review the two books—rather to show that they represent polar opposites. Frank on one side, Blüher on the other, and everthing else lies between them. Here is the whole abyss that cannot be bridged, and that can disappear only with one or other of the opponents. That Frank's masterpiece is good, more than good, hardly needs to be set forth again. It is a work of art and a hortatory text at the same time—hortatory, however, in the best sense, as we might call Lessing's *Nathan* a hortatory piece. It is no wonder, then, that Blüher sees in Frank quite the most dangerous enemy of his cause. A new Machiavelli was not exactly what Germany was lacking—but now it has one (even though in a smaller intellectual format). Hans Blüher feels himself small enough to be able to offer Germany this service. But one can hardly place Machiavelli himself in the same class as Blüher. He lived in the era in which absolutism came to maximum flower; it can still be placed to his credit that he sought not only to alert the people to their duty to the Prince, but also alerted the Prince to his duty to the people. But not Hans Blüher. What does he care about the people! "We reject all efforts toward the betterment of the condition of mankind; we aim solely at the still greater increase in the differences created by Nature"—this is Blüher's highest ethical principle. The German Nationalists and their followers will rejoice to have found so eloquent an interpreter. For in the last analysis what more could a man wish for, if he considers only himself, only his own most precious wishes and interests, than that someone come along and tell him, with every appearance of moralizing philosophy, that he does well to despise everyone else and to follow his own basest instincts. The truth is that Blüher has passed judgment on himself. Referring to Frank's book, he speaks of humans as creatures who "oscillate between animals and demigods"—and if that is the case then Leonhard Frank has the immortal merit of elevating them to demigods, while Blüher has the (much more dubious) merit of wanting to turn them into animals. This is the turning-point. As never before, humanity stands today at the crossroads. The choice must be made now—human or animal. If you choose the latter, then follow Hans Blüher—he will surely lead you to slavery, bestiality, to the swamp of the basest meanspiritedness. There let you perish—and all Blüher has to say to you is that "the State must always be some kind of slave-state," that the "grace of God is the only possible form of domination," that "the teaching of the infallibility of the Pope belongs to the most correct notions that the spirit of man has ever conceived."

To be whole is the first demand of the most elementary human existence. As little as these times permit satisfaction of the demand for moral loftiness and perfection, we still are far removed from the era in which "no truly great mind has ever been attached to the liberal or socialistic concept of the State," as Blüher thinks. It is precisely the Blüher cast of mind that mankind must make amends for. Nothing but independence of thought by each individual can bring us progress. Not the slavish submission of all to one, but the recognition by each person of the sacred vocation of his existence can rescue us from what otherwise would be certain ruin. Only truly great human beings can help this laboring century to give birth to a race that strives in social commonalty to pursue the greatest aims of all time. Belief in the goodness of the core of human nature, an unswerving attachment to the highest proponents of a free humanity are the highest ideals that we set in opposition to Blüher's servitude.

But now you might ask me, what does all this have to do with "Free German Youth"? Unfortunately, a great deal. Blüher is one of the leaders of this splintered movement. Blüher exercises extraordinary influence over the youth of Germany. That is why I can find nothing more shameless and destructive to compare with his activity. Woe unto him who abuses his influence on his fellowman; he is the most reprobate of human society. We hope that German youth will not regard Blüher's vile thoughts as the quintessence of the revolution. We hope that they will reject Blüher and not submit themselves slavishly to his influence.

Thus we have struggled with hellish potions
Upon these mountains, on these oceans,
More than with the plague itself.

To no one does this apply more than to Hans Blüher; no one more than he can refer the words of Wagner:

" I shall never envy the bird's outspread pinion"

to himself. But we, the free German youth, youth everywhere, we want to say:

I hasten forth to drink your eternal light,
The day before me, and the night behind.

## 18. COME TO YOUR SENSES!

*Die freie Zeitung* (Bern) IV (3) 1/10/20
By Walter Landauer, Heidelberg

A fissure divides the people of Germany. Other countries likewise, to be sure. But more than elsewhere Germany stands in the foreground of interest, and so also of criticism. What Goethe compressed into a few brief words has been impressed into a thousand forms by the masters of all times: "The greatest happiness of the children of earth is only their personality."

This points the way for us. It is not a way that leads with unconditional ease to solutions, but it makes possible the only solution to everything that burdens us today. *We must educate men*—this should henceforth be the content of all our actions—*who commit with their whole persons unqualifiedly for what they believe is right.* What is the point of words, if no actions follow?

Not everybody is suited to engage his energies to participate in public battles about politics and education. In accordance with their own nature many will not be fit for the task, many will not find their way to it. But there is one thing that everyone can do: to show by his deeds and his dealings which ideals he serves.

How many people are there who preach of things in their environment which correspond so little with their doings, with their own nature. In what dissonance words stand to deeds! How much cowardice lies hidden here! All those to whom this reproach applies should recognize how they add to the confusion! One rightly judges people in the first place by their actions. Otherwise how cheap and easy are ideals! It is often difficult indeed to be equal to this demand—but it is an inescapable obligation. It is here that the most devastating disease of our time festers.

The absence of unequivocal characters paralyzed the revolution, and inhibits every further development. Germany possesses thoughtful scholars, many poets, imaginative artists—*but it has no complete individuals.* The weightiest accusation that can be brought, yet it must be brought—and repeatedly—is not against individuals, but must be directed against the great majority of the nation.

Because people arbitrarily separate their politics from all the rest of their lives, instead of linking them intimately, they cut off—quite often unconsciously—their work from their lives. Let me mention one exemplar: a Russian, Leo Tolstoy, stands gloriously before us, his whole life dedicated above all to the aspiration to bring words and deeds into accord. We love and venerate him for this above all others.

One can turn only with revulsion from the life of our own days. The words of Charles Fourier are applicable today as never before: "It is the deceitful state of commerce that is the cause of the falsity of all our relations." Our entire business life is decayed, corrupt, and pestilential. The consumer has long since been driven into silence. Tradesmen and manufacturers compete in industry and agriculture to bring about the all but unavoidable collapse. One preaches to deaf ears if one does not speak of profits, convenience, and pleasure. We listen to these fine words, repeat them complaisantly, or put up with them. But the consequences? Nobody wants to hear about it.

In your business affairs—here even more than in other matters—you can show that it is more important to you to raise up a new mankind. Each individual, whatever his position, is called upon to participate in the betterment of man. Whoever takes the trouble to examine his own actions with reference to the general good helps to reestablish solid ground, to fill in the swamp into which everything threatens to sink. The primary aim is not the improvement of others, but for each of us himself to come to his senses. That is where the effect on others begins.

Know this, O Germany: this is the only remedy that remains to you. Individualism does not mean that everyone pursues his own instincts and interests, and only those, lawlessly and ungoverned by rule in accordance with their own judgment—that brings them to the threshold of the basest attitudes. Individualism is the striving by each person through the development of his responsibility to contribute to the welfare of the collective. Develop character, become whole persons—only then can our efforts against the miseries of our age prove fruitful.

## 19. ON GERMAN PACIFISM

*Die freie Zeitung* (Bern) IV (9) 1/31/20
By Walter Landauer, Heidelberg

Pacifism plays a peculiar role in spiritual life today. It has frequently become a farce, not on account of its ideas, but because of its representatives. During the war much was said about the collapse of pacifism, and during the revolution pacifist circles rejoiced at its victory. The one is as erroneous as the other. Pacifism is a teaching and a faith. Teaching in the sense of Buddha, faith in the sense of Christ. Pacifism must be lived, *and be given expression in men's lives.*

Because there are so very few people who are capable of taking seriously an idea by which they want and are able to live, is the reason there are so few pacifists. For the true pacifist, whose belief springs from heartfelt conviction and who does not propagate pacifism for some utilitarian reason or another, the war only succeeded in confirming what he already knew; but the German revolution has taught him how far mankind still is from conduct free of violence.

It is not a question today how many people in Germany *call themselves* pacifists. Only those whose pacifism is united with faith in mankind, who are able to avoid hatred, who are impelled by love; those for whom it is the supreme commandment and principle of their conduct; those who cannot live without it, and derive from it the will and drive to live; only such people can possibly spread the true spirit of peace. What passes today for pacifism *is as virtually washed-out as the nearly proverbial "German democracy"*, the spurious imitation of great thoughts, mean-spirited perfunctory pantomime, the Christianity of the churches.

No wonder, then, that pacifists lack all inner unity. They differ widely from one another in tactics and politics; they are to be found in the most varied parties. Not long ago I said in this column "that every party, regardless of its name, aims at dictatorship and suppression," and from this it follows that no true pacifist can belong to any of the German political parties.

Someone coined the phrase: "The radical pacifist sits in the parliament of man to the left of the Bolshevik." Let us understand this dictum correctly. Marx, the genie of the socialist parties, acknowledged the curse of the violence of the bourgeoisie, and sought to transfer this violence to the *proletariat*; he bought this truth with a much worse lie. But those who wish rightfully to call themselves pacifists must recognize the curse of *all forms* of violence and act accordingly. For them there is no more authority outside their pacifist beliefs.

*Sub specie aeternitatis* the Marxism of the *Communist Manifesto* is nothing less than radical—in opposition to it, pacifism rejects all previous history. Only someone who has grasped this is able to understand how utterly difficult the task is, especially for the German pacifist. It is not insoluble, provided that a disposition toward peace is called into life. Recognizing this is for nobody as necessary as for most German pacifists.

## 20. THE MATADORS OF THE "FREE GERMAN" YOUTH MOVEMENT

*Die freie Zeitung* (Bern) IV (11) 2/7/20
by Walter Landauer, Heidelberg

Let us accept as preamble Goethe's saying: "Youth models itself on youth." If we still have hope in Germany today, it is above all else our hope for youth. The masses of those whose roots lie in the past century are all too much under the influence of the then prevailing system to be able to liberate themselves from it at all. But if the concept of youth already includes in its essence that it is revolutionary, that it doubts and shatters old traditions, searches for and strives to attain new pathways and higher goals, then it is truly the case that *the most depressing characteristic of the German revolution is that its youth has so far taken so small a part in it.*

Any observer of the German youth movement during the last few years could cherish the hope that it would adopt the new ideas of a revolution and carry them forward. How little of all that has occurred in reality! Rödiger's words from his bonfire speech at the Wartburg Festival in 1817[1] are as valid now as then: "I tell you, young men of Germany, and oh!—— that I could thunder it into your hearts: this is our mission!" One leading thought runs from the founding of the fraternities (*Burschenschaften*) to the student ramblers (*Wandervögel*), from the "free academic army" to the "free German youth." It was expressed by Gustav Wyneken in these words: "Youth, which has hitherto been an appanage of the older generation and shut out of public life, reduced to the passive role of learning, to a vacuous playful sociability, has begun to think for itself. They are attempting to shape their own lives..."

It was out of this thought that the first *Wandervögel* movement grew 23 years ago; it rapidly gained recruits and rambling groups sprang up everywhere which, however, from their inception were widely suspected to be closely allied to official military endeavors ("Youth Army"). In 1913 the first "Free German Youth Day" assembled on the Hoher Meissner near Kassel. It was when this was established that the inner schism in the whole movement became completely clear. *Gottfried Traub, Ferdinand Avenarius, Ahlborn, and Wyneken* all gave speeches on this occasion—and these names alone would even then have enabled the prediction that the movement had no unified or long life expectancy. A powerful spiritual bond, some revolutionary inner experience, would have had to be present to overcome all the existing contradictions if it was to lead to any objective unity. And that is just what was lacking.

Now is the time to assess the movement, to ask whether it still has any meaning for us in the future, whether we can still expect anything of it, or whether we must oppose it. The difficulty alone of giving a critical account of it at all demonstrates the confusion that surrounds the alleged impulse for this association. A. Grabowsky has recently published under the title *The Free German Youth Movement* a collection of essays by leading personalities of the free German youth. Leaving to one side the fact that the selection is excessively dominated by aspects ("Diplomacy," "Politics") designed to give a false overall impression of the movement, this very selection also suffices to display the intractable discrepancy clearly enough. Grabowsky himself, the only nonmember of the Free German Movement who is included—and he is surely

[1] Translator's note: The Wartburg Festival took place in 1818

not to be suspected of wild rebellion—utters the best words for the youth to whom all these essays are directed. Grabowsky believes that "a religious devotion" unites all these people "in the spirit of a purified Christianity prepared for new deeds." If that were the case, how joyful we should be! But it is here, on the contrary, that our criticism enters. Implacably, desperately, I would almost be inclined to say. Not just because in these voices of the "leading personalities" no sort of unifying theme is to be found; it is nationalism[1] and materialism of the very worst sort that dominates. But that one can speak in this of "religious devotion" is simply a farce.

Among the leadership of the Free German Youth the most prominent personality by far seems to me to be Gustav Wyneken. Wyneken, who founded the "Free School District of Wickersdorf," and who is now heading it again, unquestionably stresses the value of education in an excessively onesided manner. To be sure, education is of immeasurable importance, and one would not in any way want to dispute that Wyneken has achieved much to bring about its reformation, but it is not only erroneous but misleading to believe that education for pedagogy's sake is the alpha and omega of human social life.

Wyneken is fighting against the acceptance of authority, but personally exercises—perhaps as a result of his unconscious power of suggestion—so very authoritative an effect (and unfortunately not in the best sense of the word), that his educational talents are thereby neutralized. In his writings he so little acknowledges any specific *political* convictions that it would be difficult to establish what he thinks about *other* than purely pedagogical questions, if we did not know that between 1914 and 1918 there was a time when he *endorsed the German war.* Wyneken is hardly the man to build anything new from the ground up. Sentences like the following: "What a consecration the war is for you!" "To *be able to endorse this state* (of war) *without reservation and with joy* seems to me the fair privilege of youth." "Since it is the Kaiser, like his triumphant grandfather, *that God alone wishes to honor,* he undoubtedly expresses the feelings of the widest circles of the people." Such utterances do not lag in any respect behind the Manifesto of the 93, or behind Pastor Traub.

I would not disinter these samples from the past if the Berlin paper *Aktion* had not reported that as late as July 1918, Wyneken "expressly affirmed" that he stood behind the meaning of these words now as then! That might be an occasion for thought when we are dealing with this "most revolutionary" of youth leaders.[2]

---

[1] Translator's note: text reads "rationalism"; almost certainly a misprint

[2] He has recently followed Wilamowitz-Moellendorf to Switzerland, and his lecture-series in Basel-Zurich-Bern is said to have made an "impression." [Ed.].

I pass rapidly over the throng of other leaders, who are sadly all too lacking in personality, whether they are reactionary like Prof. Messer or doctrinaire like Dr. Ahlborn, in order to come to the other extreme. That is Hans Blüher. Beyond his writings on eroticism, Blüher has so far chiefly devoted his efforts to the youth movement; along with Wyneken he has certainly best understood how to make himself influential. Even on superficial consideration it is extraordinarily noticeable in how similar a fashion a volatile instability dominates in his case too. If you can get past the importunate, contrived and unbelievably conceited style of his writing, and press on in his works, you will be amazed how frequently totally irreconcilable thoughts are uttered there in a single breath. Just to give one example: in one of his essays Blüher supports his views by citing the thoughts of Martin Buber while insistently commending to youth Gustav Landauer's *Call to Socialism*. In the preceding sentence he speaks of the unforgettable services of the Social Democratic majority Party, and a few sentences later he says: "The whole human nature of leftist culture is in the last analysis completely pointless."

I will now assemble here succinctly just a few passages from Blüher's writings, only from those published in the year 1919. "For after all any war between strong peoples is a great affair. That they [German youth] *have stumbled into the revolution,* that is much worse ... Something like that should not have happened." "So shrug them off, these revolutions!" *"There's no such thing as love of mankind!"* "The only things equivalent to a hero's death are *the sacrament of last rites* and a willing death in *a condition of triumphant perfection*. "To be in God's grace is the sole possible form of domination."

Enough! The majority of readers will surely have long ago said: "Methinks I hear the utterance of a chorus of a hundred thousand fools." Yet it is necessary just once to realize what sort of people today's leaders of youth in Germany are. I particularly emphasize what *freedom* means for the Free German Youth Movement in contrast to other tendencies. Anyone who knows German conditions knows what a range of meanings this ominous word can take on. But even if not all the leaders of the "Free Germans", like Blüher, follow a stereotype, "then an utter contradiction subsists, as mysterious for sages as for fools," and all are lacking, in sum and substance, the all-encompassing spirit that impels to new goals, the overflowing enthusiasm, the convinced resolution to bring about a spiritually renewed human society.

Even worse than the shortage of leaders is that youth allows itself to be ensnared by disguised phrasemakers, that it is still unable today to participate constructively in shaping its own future, but instead slips back into a shallow past. Before it is for long too late—and soon it will be—youth must acknowledge that this is for them *a matter of* being or nothing, that they now must form their resolve or no longer be able to do so. What is needed is a decisively defiant release from all previous superficial complacency, to gain sight of a goal and unswervingly to pursue it. Honesty and logic are required. I join Wienbarg in saying, "Things must be different: that should be the feeling that dominates everyone; that is our calling: the powerful echo of that feeling."

## 21. A SPIRITUAL LEADER OF GERMAN YOUTH

*Erkenntnis und Befreiung* (Vienna) II (11) 2/8-14/20

Among the Shachsevs of the Moghan Steppe blood feuds always end with marriage between the hostile parties. Kropotkin reports this.

What a spotlight this casts on our times! What darkness surrounds us! More than a year of revolution and still the fetters are no more relaxed. Fetters by which the people are held captive; fetters which they themselves forge ever anew. No wonder then, that even individuals who perceive the goals of mankind begin to postpone the path to their realization to the next generation or the one after that.

No, nothing is achieved by those means! How should the next generation be able to function better and to attain freedom, if its *education* is at the mercy of this one? Eventually the total spiritual transformation must come about; each day that it is delayed is lost to mankind.

Mankind is entoiled in an ever-increasing morass—to sink and suffocate there, *or* to be liberated. These are the alternatives.

What will be displayed here is what it is that threatens youth today. The "Free German Youth" movement has for years embraced all classes of German society. Diverse tendencies have emerged within it for a long time which understandably have been struggling against one another now more than ever. Blüher is the leader of one side; his influence is not inconsiderable, and his ability to inflame the minds of men with his words is astonishingly powerful.

We have lately suffered the misfortune of receiving an offspring of his spirit (*In Medias Res* is its name, and it is further entitled *Fundamental Observations about Mankind*). It would surely not be worth cleaning the Augean stables just for the sake of looking at them, if it were not for the fact that something of significance for these times were at issue. "In no wise has mankind made any progress." This is Blüher's loftiest insight.

With crass misunderstanding, projecting himself as a follower of Schopenhauer, Blüher casts him in the role of promoter and prophet of the most extreme despotism! For him the Prince is the spirit who hovers above the waters, he is the "supreme repository of the essence of his people." One link is fitted into the next—and the shackle of slavery intended by God is complete. By their very nature, peoples hate one another, politics can only be conservative, God's grace is the sole avenue, the unchallengeable infallibility of the Pope, the wickedness of all the Jews—all this is self-evident for Blüher—in a word, according to him, one must "stand to the right of the Junkers."

Briefly, according to Blüher, there are "born nations of slaves," and he desires *that the Germans remain such a one, and become even more so.*

We have nothing to fear from the spirit of Blüher, not even from his sophistries—but anyone aware how readily susceptible people are to slogans, no matter how senseless and objectionable they may be, knows also the grave danger posed by any prophet of power. The curse of the socialist parties is their name. That can be the only reason that many people believe today that socialism has now demonstrated that it is neither able nor entitled to create a new way of life. That is the opinion of all who have never possessed the will and capacity to master truly socialist ideas, including the "Germanic gentleman," Hans Blüher.

Even though we have picked out the most extreme of the [Free German] Youth leaders, yet there are still enough of the others who systematically undermine the narrow defile into a better age with their similar lofty (and empty) sounding words. The doctrine that we must above all infer from this once again is that our efforts must at all times be directed to stimulate the creation and revival afresh of thought and will in every individual. Our generation was born to renew mankind. That may or may nor sound presumptuous—but true creation can be attained only by those who truly seek in their inmost being to be creators. And it is high time to recall the words of Charlotte Corday:[1]

"We work with greater energy toward our own downfall than we have applied to the conquest of liberty! Oh, ye French, only a little time yet and nothing more will remain to you than the memory of your existence."

*Walter Landauer.*

[22, 23, 24: G. Landauer's murderers acquitted; review of Lehmann; The Internationale—Germany omitted.]

[1] In her *Summons to Law-abiding and Freedom-loving Frenchmen* of July 12 1793, written the day before the murder of Marat, five days before her execution.

## 25. FROM GERMANY

*Der Aufbau* (Zurich) II (45) 10/8/20

### *Die Bauernzeitung*[1]

A new paper has been founded in Germany which, in contrast to a thousand other new publications can command respect in the widest circles from its inception. It is the *Bauernzeitung,* thus directed mainly to farmers, but important also to all people who participate in building the economy of Germany and of the whole world. This journal deserves consideration because of its program, which will be described here in a few words. The *Bauernzeitung* will work intensively towards the establishment of a new social order. To this end it proceeds from the concept that this can be done only by starting at the source of all human life, the soil. The capitalist economy has bred its own downfall. Highly intense competition has led to an insufferable mistrust among people and, indeed, among nations. Only this cast of mind could have led to the World War which means the ruin of Germany unless this pattern is fundamentally altered. So far, the revolution has produced hardly anything creative; it proved sterile throughout the squabbles of the parties in the struggle for governmental power. Thus, from the beginning, the *Bauernzeitung* established the basic principle that it would serve no political party or industrial union; rather that, independent of any party line, it would demonstrate in its coverage the strictest adherence to the truth.

Business and industry move from crisis to crisis towards collapse. Under present conditions farming cannot resist this development and it will, if it does not swiftly realize the danger, share the same fate. The only future for farming, and essentially for humanity, lies in a general working community which is built, not upon envy, hatred, and competition, but upon mutual assistance and love—we have in mind *cooperatives.*

People who live with each other and work together every day, alienated from each other in their innermost beliefs through competition, must join together in communities with similar interests. The natural basis for this process is the village, and the next step is the local commune. With its own management it protects the common interests of the group it unites—the commune. District organizations consisting only of representatives of the communal cooperatives are the link between city and country. They will exchange their produce with cooperatives in the city, and through this process meet the needs of both. Middlemen, the real abscess of capitalism, must disappear. The existing federations and cooperatives, especially those which involve coerced participation, are based entirely on capitalist principles and differ in no way

---

[1] The *Bauernzeitung*: published by Freier Bund in Ueberlingen am Bodensee has been appearing Saturdays since September 18; monthly cost 1.50 Marks; postal delivery.

from other business enterprises. If these are treated as the norm it is not surprising that farmers are mistrustful of cooperatives. Therefore, the *Bauernzeitung* has repeatedly to point out the nature of the true cooperative. The beginnings of real cooperatives already exist— they show what communal work can accomplish; and their example will stimulate interest. The general self-administration of the cooperatives has to spread from the country. Naturally the main foundation for the maintenance of the spirit of joint work and mutual assistance must first be established. That is the school. Hitherto schools have been subordinated to the purposes of the State whose only purpose was to educate compliant tools for its own objectives. This must change fundamentally if cooperatives are to flourish. We need free people who are as conscious of their duty towards others as they are of their own self-esteem. The cooperative has to be a community of people freely united on the basis of common goals. On this point too the *Bauernzeitung* has to start in the country. It will give advice to farmers and transmit information important to their work. This is not enough, of course. It therefore proposes that rural schools be set up everywhere in which people will be prepared in the spirit of brotherly cooperation in agriculture. The spirit from which the guilds once arose, when a free peasantry managed their own commons, must again be awakened and form the basis for a new community of men.

As can be seen, the *Bauernzeitung* seeks to participate in the realization of a powerful movement. Most important at the outset is its involvement in founding local rural cooperatives, and in providing support in word and deed for farmers by advising them in areas where they want to combine themselves into a cooperative. The cooperative itself is supported by a union of people who are pursuing the same goal; *all of the profits benefit the movement (especially educational objectives).* As a motto these words will serve: a precondition for a free humanity is the cooperative free peasantry.

*Walter Landauer*, Heidelberg.

## 26. A MIRROR TO THE WORLD

*Der Aufbau* (Zurich) II (47 ) 10/22/20

***Prefatory remark.*** With this contribution from Walter Landauer we would like finally to inaugurate a long-planned series. Among the most important tasks of a real educational enterprise is to find a way through the chaos of literary phenomena that furnishes help and draws attention to the good and valuable. We very much hope that this part of our paper will attract contributions from many readers and friends who are otherwise not in a position to participate, each of them from his own area of interest and activity. We expressly note in this connection that we do not wish to limit ourselves to original writings.

*The Editors.*

"Free by my own action and true to the
fatherland, what duty bids is sufficient for me."

Voltaire's voice—it was Charlotte Corday who uttered it before she climbed up to the bloody apparatus of the revolution. Voltaire's spirit, Charlotte Corday's courage, are both needed today to tell the truth to the German people. The most difficult duty for revolutionaries is, during a revolution, to have to say to the people: despite the revolution you have not overcome the domination of the lie; you proved unequal to the task of establishing new values of lofty morality instead of what was overthrown. There is no revolution where the lie remains a lie, where the old spirit lives on beneath a new dress, where violence is driven out by violence. Once again we hear an all too rare voice of truth powerfully raised. *My Struggle against Militaristic and Nationalistic Germany* (Verlag Friede durch Recht, Stuttgart, Werfmershalde 18.), Friedrich W. Foerster's enthusiasm, the benevolent and beneficent call to brotherly love, does not display the tortured mien of anxiously doubting concern, when Foerster raises his voice in warning, admonition, and adjuration: "I am a German, and am fearful for Germany; I am apprehensive about Germany's rescue from the Prussian spirit, and be sure of one thing: you will not get rid of me." We have enough revolutionaries; we need people of revolutionary spirit. Foerster is one of them.

Already in its organization the book shows the way the spirit of the German people must go if it is not to founder on its own guilt. Death to lies; long live the truth! Thus every renewal today must begin with the question of war guilt. Not with bureaucratic files and an unbroken chain of evidence—no, "this materialism of files and facts has only the appearance of a solid foundation. In reality the spirit resides here along with the letter." I felt this already in 1915 in *J'accuse*, Grelling's provocative book of accusations, that the most convincing proofs of guilt lose their force at the point where an elevated will for moral renewal is lacking both in the accuser and the accused. From sociological and psychological perspectives Foerster makes the whole tragic development of Germany as the instigator of the war pass before us in review. In these scrutinies one must always distinguish between general and specific war guilt. On the general guilt Foerster's conclusion reads as follows: "And this seems to me to be the essence of Germany's guilt for the war: We drew the ultimate and harshest conclusions from the dissolution of the European *civitas humana*, from the doctrine of the dominating force of mere egotism, from the traditions of violence of Machiavellianism; we became the logicians and systematists of the military state built upon the geopolitical right of might; we consummated the incarnation of the evil spirit of the history of modern nations."The problem of Alsace, our treatment of the population of which makes it morally fully justified that it should be "taken from us" again, the shamefully ugly stance of Germany in the Morocco conflict, the alienation from England for which Germany is responsible (on which Foerster, in the boundless evenhandedness which he takes for granted, also treats of England's sins of omission), the Slavic conflict, the outrageous conduct at the Hague Conferences—all this comes to life in a numbingly powerful accusation. The means becomes the end for these new Germans, the purest morals are abused to encompass a categorical submission to categorical and organized immorality. "Be quite clear about this, Germans, you are cast out from the community of nations if you do not display your bitterness against a system that has made you the executioner of your neighbor, and in the end has corrupted you… The world will not receive you until you become a human being."

Who are the guilty parties? The materialists of power politics, the ideologists of power romanticism, the casual fellow-travelers and yesmen. "At the level of elementary school teachers these worlds divided," says Foerster. On the same plane as this outlook was the way we ran the war. It separated Germany still further from human society. The violation of Belgium's neutrality, the fanatical lust for destruction, barbaric gas and air attacks, unwarranted transportation, the reckless rejection of every possibility of accepting peace—these cannot be expunged overnight from the memory of those affected. "For the German people there is now no more important world-political education than practice in empathetic understanding of the motives, feelings and thoughts of the opponents, and from that standpoint to see one's own actions and what one has condoned in a new light."

Kierkegaard's sobriety meets our need: to come so close in our understanding and acknowledgement of ourselves that all that understanding is translated into action. Only in this wise can redemption arise out of so profound a guilt. Only then can we approach true politics—the art of winning confidence. "Let there be an end of the State governed by violence; only out of some new reciprocity in mutual understanding, mutual advancement and living and letting live can a new social life begin for us." It is in this manner that reconciliation also must come with France, which is for us, as Boutroux said in 1914, not a contrary, but a complementary. The spiritual and moral demobilization of Germany is the precondition for the success of military demobilization. Beyond that come then our position on the territorial enclaves, on the German-Austrian question, on the development of world trade. But a major problem is the *new German internal politics.* Today we stand in the midst of a total Prussianization of Germany. The spirit of the people has not yet liberated itself from centralism and dictatorship: "This evil should deliver one last great service to the good, and then withdraw. This dedication to contradictory principles, however, has lamed the whole thinking and aspirations of the revolutionaries, condemning them to sterility." The Republic has so far brought us nothing but an intensified Prussianization—an enforced economy together with moral anarchy—but what we need is self-governance through cooperatives, in towns and districts. Not nationalization, but denationalization is the demand of the day. Socialism must progress from the revolutionary to the conciliatory deed. "Everything depends, then, on the German's conversion from a mechanical synthesis to a synthesis reached by humane means."

Foerster's book: a mirror to the Germans, a mirror to the world. So much that is repellent blazes forth to us from it, but it only reflects faithfully the truth about the events of our day. The spirit, the sublime spirit of true revolution that is uttered in Foerster's words, is not alive in Germany. It will be—or Germany will founder.

The burden piles up,
With increasing guilt,
Of the woe of the world.

*Walter Landauer*, Heidelberg.

[27 G. Landauer and Bavarian revolution omitted.]

## 28. MUNICH IMPRESSIONS

*Der Aufbau* (Zurich) II (52) 11/26/ 1920

Why first impressions so often capture what is important, substantive, and typical is familiar enough. Anyone who steps from other circumstances into some new living environment is at once struck by the different expression of those circumstances. If one's talent for observation is not equal to the task, the law of inertia comes to its aid. Perhaps the following lines, even though fleetingly gained during a brief stay, will capture some significant features of the characteristics of the life of a city—to the degree that it can be differentiated from that of other cities. As mere sketches they make no claim to comprehensiveness.

Munich is a city in which one can hear, in the street and in the crowded streetcars, that people greet each other with "Hail, Aryan." These words vividly reminded me of the greeting for so long preferred during the war: "Hail and Victory." How much filth comes to light in two such short words! Will the fate of the first be like that of the second? We expect so, in fear and hope. The hunt for Caesar's crown foundered on the outraged resistance of the whole earth; yet a hateful exclusivity cannot be overcome from without, but must be conquered from within, if it is not to act as a deadly poison, like a medium that hastens the miasma of dissolution.

Another vignette. In the busiest area of the city—near the railroad station—a whole block is cordoned off for hours. Battalions of army detachments, acting in rising hopes of the return of times that we had believed at the beginning of the revolution to be gone forever, exercise unrestrained violence. Passers-by, of whom one inquires as to the meaning of this activity, answer brusquely as though amazed at one's ignorance, as they hurry by "It's a raid!" Trucks and prison vehicles spell each other in front of the site selected on this occasion, are heavily loaded with people and race away. Hardly any of them is successful, even with a hundred protestations of their innocence at heart (just that is often their greatest guilt!), in escaping this furious nemesis. In the scurrying bustle of the streets of the metropolis the number of those is small who observe these events seriously, depressed and silent. In the psyche of the masses it would seem that these constantly repeated events barely rise above the threshold of consciousness. Yet even though this outrageously vile tyranny is for the present no longer perceived in all its baseness, still it is the case that the hopes attached to it by the reactionaries are fundamentally a failure. As uncertain as our future still is, one thing is nonetheless certain, and that is that medieval autocracy is forever behind us—at least if it comes from the Right.

Troops of the much vaunted, but never seriously opposed security services—*Orgesch!*—identified by armbands, march armed through the streets, and assemble for exercises and solemn shooting meets. These supporters of reaction are avidly engaged in bringing the workers beneath the yoke once again! The citizen who still has not grasped this knows nothing of the connection with the demand (just the same wholly unjustified) for the destruction of the Diesel motors.[1]

Even more than in other cities we see in Munich another typical aspect of our life today. Delicatessen shops of all kinds have undergone an unprecedented rise in prices. That they are much more frequented than any other kind of shop is deeply rooted in the developments of our times. The progressive rise in

[1] Cf. in this connection the article *Orgesch and Diesel Motors* under the heading "From newspapers, journals and books."

prices, the steady deflation of money has brought about the ever-increading mistrust of paper money. The fact that by far the greatest proportion of the nation is living better than before is not, of course, attributable to better sources of income—salaries and wages have almost without exception not kept pace with inflation—but *is to be explained that nothing is spent save to satisfy the most animal needs.* Carried to this extreme we find here a horrifying picture of the decline of our culture. In every class of society an outward superficiality dominates; people display a frivolous lack of concern that seems to precipitate us into the whirl of a dance of death.

The most striking characteristic in Germany—like all these shadows a few degrees darker in Munich—is the spirit-crushing *domination of the bureaucracy.* "How is patriotism possible, given the total dependence of German officials upon the government," we read in an article on *Bureaucracy and Officialdom in Germany*, dating from 1844. Today that relationship is reversed and we might ask: how is a German government possible given its total dependence on the arbitrariness (read: patriotism!) of the officials. And Freiherr von Stein once said: "We are governed by salaried, booklearned bureaucrats with neither interest nor property—that's all right, so far as it goes." How much further should it go? How much more misery must we endure before the sins of the old system are atoned for by actions springing from below, until from the lessons of 1914 the will blossoms forth to embrace truth and humanity, until the worship of violence falls silent in face of an impulse to community? And yet this thrice-accursed bureaucracy (which—and I give only a single example—requires in Munich that every "non-Bavarian" who stays for more than 24 hours must report to the police on pain of a substantial prison sentence), cannot prevent me even in this desolate present and the boundless misery of our coming days from saying, perhaps somewhat sadly, with Hermann Hesse:

And in spite of every pain I am still
In love with this crazy world!

Walter Landauer, Heidelberg

[29 Review of Plenge omitted.]

## 30. HALLE, KASSEL, HANNOVER

*Der Aufbau.* (Zurich) II (55) 12/17/20

A whole string of German political parties held their plenary meetings during the last few weeks. Annual accounts were presented, and the action agenda was set out. Little that was new showed up in one or the other of the party meetings of the Communist Party, the Communist Workers Party, and the German People's Party. They contented themselves with wallowing in the old swamp of slogans, and it is all the easier for me to say nothing about them since none of them had anything decisive to say about the future of Germany. Three other party meetings will nonetheless occupy us briefly today. Each of them casts some light on Germany's present-day existence.

Since people's interest is most readily excited by noise-making, the meeting of the Independent Party in Halle[1] attracted particular attention. The internal tension that had been building up in the party for months reached its climax and led to the expected result: the defection of the majority to the communists. From this party meeting, as from that of the Social Democrats in Kassel, I will pick some things out that could perhaps point to the dawning of new perspectives. After all, that is all we have been searching for with increasing longing.

Let us therefore enter this quarrelsome circle with no expectation of spiritual returns. For once again it was apparent in the crudest fashion how frightfully barren, devoid of intellect and spirit Communist Party politics is, no matter where we encounter it. And for the most part it was from start to finish the politics of the Communist Party of Germany that was being played out. Read just half a dozen of the legions of agitation tracts, and I could think of nothing new to add to their doctrines. In accordance with an easily understood law it is always the case, where a minority is desperately protecting itself against oppression, that more humane tones are struck. Thus it came about in Halle that from the benches on the right some significant words were heard that went beyond the party bromides. For example, Hilferding: "We must be prepared for all eventualities, so as to exploit them. Therefore it is necessary that the revolutionary parties do not compete disgracefully with each other in radicalism; that would lower the *proletariat* in the eyes of the bourgeoisie; that would serve only the interests of the counterrevolution." "We need a policy that does not entail dictatorship over the masses of the people, but one that wins them over so as to attain a Social-Democratic democracy with their help." "These constant provocations simply open the door to every political adventurer," this last a

[1] I will return to this in another connection.

pointed message to Zinoviev and Moscow. I find the dramatic climax of the meeting, not in the inflammatory speech of Zinoviev, not in the resultant rebuttals, nor in the ensuing splintering of the party, but in the words of the Russian Socialist Martov: "Nothing has wounded me so much, and nothing reveals more clearly the state of the Russian revolution, than the question posed to me here in Germany: what would they do to me in Russia if I speak here in Germany against Zinoviev?" How few were those who felt all the weight of the inexpiable guilt of this frightful accusation! The conclusion seems to me indubitably correct that Hilferding drew following the splintering of the party meeting about the politics of the Left: "What do the Communists live on in Western Europe? They live off that painful demoralization, off the overthrow of all values that has been produced by the war; they live off the degradation of mankind brought about by the dreadful mass murder of people in these last years." But we can scarcely believe in his words, given our previous experience, when he continues: "We live, party comrades, we live on the rise of rationality while they die of their expiring illusions."

The party meeting of the Social Democrats ran a calmer course in Kassel. Since, when they assembled, they were already convinced that Moscow's conditions were not negotiable, the meeting mainly concerned itself with separate issues of economic, parliamentary, and governmental activities. That said, plainly nothing of moment was addressed.

In his keynote speech Wels summarized the economic and cultural problems of the world in these words: "How do peoples and states, by mutual promotion of productivity, restore as adequately as possible the destruction and damage caused by the war." There can be no doubt that if this thought would enter the psyche of the German people completely, things would be substantially better for us. To be sure, there is an enormous amount that needs doing here. But that is not the program of the workers' movement; nor that of socialism. *And this was not talked about in Kassel at all!* An external high point—I did not find an internal one—was presented in the confrontation between Wissell and Robert Schmidt (both former Economics Ministers of the Reich) concerning economic policy. One advanced a planned economy, the other the free market, as the solution. Wissel says: "Capitalism and commerce have been able to indulge themselves, as though there had been no revolution in Germany." "With relentless rigor, even with brutal severity, all consumption that is not absolutely necessary must be suppressed." Schmidt proclaims in opposition: "Let everything go that we can dispense with; our foreign trade must get underway!" Once again, as is nearly always the case when people quarrel, the meaning of the words renders the discussion illusory from the start. For why would it naturally follow that both programs should exclude each other like fire and water—when only too readily they could and should be combined! A lot of time was taken up with the preparatory discussions of a new party program, for the elaboration of which a commission was established. Once again, it was an outsider—Eduard Bernstein—who spoke words of warning at the right moment: "We must only beware of Marxist scholasticism. You can prove anything with Marx, because at different times and under different circumstances he expressed different views in different language."

The main task of the party meeting thus lay in the area of reform. And it should be reiterated that the most urgent demands were presented as aims of lofty ethical and cultural significance, which it is truly worth committing oneself to. I mention only: suppression of brothels, subordination of military power to civilian authority, advancement of the cooperative movement, improvement of agricultural education, the most rapid nationalization of fertilizer plants, etc.

If we look beyond generalities about reforms and party interests for the results of both party meetings, however, we scarcely find, even with the utmost goodwill, any reference to what we and the German worker most urgently need: the definitive renunciation of doctrinaire belief, of blind acceptance of authoritarianism and petty party interests. An unqualified commitment to the interests common to all workers. What Stöcker, one of the most rabid of the Communist Party faithful, expressed in Halle: "We are of the opinion that we do not split or splinter, do not expel; we unite, educate, enlighten—that is the task of a revolutionary party." Truly, no one of our parties has yet elevated this to its highest principle. And for that reason none of them seems to me really revolutionary. What developed without friction and organically in England, in the unification of all factions in the Labour Party so as to represent all fundamental objectives—that seems to be impossible in Germany. Yet it is only in this direction that it is possible to work fruitfully for the worker; and our entire efforts must be pointed to this end.

Finally, Hannover. There is no point in reporting on the party meeting of the German Nationalists, who once again formulated their program in their old fashion: "Prussia is our ultimate objective; from Prussia we aim to conquer the Empire." "When we Germans are truly free we shall someday succeed in reshaping the external world in accordance with *that freedom,*" a phrase we could ignore, if it did not hint at a strategy. The party of the "glorious Traub," as Graefe has expressed it, has sufficiently understood the times to know that as constituted so far, it has no future. Thus it seeks to lure farmers and workers to its numbers. On the one hand they say: "Only our party does anything for agriculture," and on the other: "Social questions are dearest to our hearts." Even if the danger is not great that workers become enamored of the "law and order" state that has become a reality in Bavaria, still the splitting of the Socialist parties advances its [the State's] demagogical tendencies.

One could ask, then, whether Halle, Kassel, Hannover signify some sort of sequence in a linear development. Why I regard that as out of the question has already been hinted at above. The ideal of the German Nationalists lies in tatters on the ground forever, yet the workers' parties have not yet succeeded in finding the common path that alone would lead to beneficial results. Working for the realization of socialism will only be possible for them when they have destroyed all megalomania and lust for authority, have rejected every mean-spirited dogma, when they place only the highest common interest of all productive people in the foreground of their actions. Until then, we must wait, warn, prepare.

Walter Landauer, Heidelberg

## 31. ON THE PROBLEM OF VIOLENCE

*Der Aufbau* (Zurich) II (56) 12/24/20

In Numbers 35 and 50 of the *Aufbau* the problem of violence was addressed. I would like to take a position here in a few words on those articles.

Now it seems to me very useful in the first place to extract some statements from the proceedings of the USPD (*Unabhängige Sozialdemokratische Partei Deutschlands*—German Independent Socialist Party) in Halle, which show how the thought of the use of violence is imagined in different minds. On the Left, Stöcker says: "The Party will have to accomodate itself to the battle for the dictatorship of the *proletariat* ... The dictatorship of the *proletariat* is the exercise of violent dominion over the bourgeoisie ... At the Leipzig Party meeting Comrades Ledebour and Hilferding rejected terrorism as immoral. But we know that we can conquer the bourgeoisie only by the use of the same means that the bourgeoisie directs against the *proletariat*." And, in the same fashion, Däumig says: "How often have we not heard in recent times in our meetings: We do not want civil war. But in this connection it has nothing to do with what we want, rather we must consciously prepare ourselves for civil war." Against this view, speaking from the right wing of the left half of the party session (please excuse this unavoidably complicated expression!), Crispien enunciates the view: "Matters are not the way the Communists see them in equating terror and violence. We all know that the final battle cannot be conducted without the use of violent means. But we do not know, and cannot foresee, what means and to what degree violence will have to be employed. We shall make use on each occasion of the means dictated by the situation ... We declare that we acknowledge the dictatorship of the *proletariat* as a temporary phase, as the transition from capitalism to socialism. What it signifies is the deliberate, conscious, planned exercise of political power governed by socialist morality. That is what we understand by the dictatorship of the *proletariat*." And finally, from the Right, we hear the voice of Dittmann: "We must oppose war for reasons based on our socialist convictions ... We will not turn anyone into a socialist by violence."

What we see here now in authentic fashion are the most diverse conceptions of violence as they prevailed in the now riven USP [Independent Socialist Party]. I am compelled to agree unconditionally with Hubacher when he says that the approach of the Left is more coherent than that of most Social Democrats. In taking a position on problems of such decisive importance coherence of the most rigorous kind is necessary. Let us therefore be on our guard above all against mutual misunderstanding! The word "violence" has many meanings. Nothing will ever persuade me to believe that the success (*sub specie aeternitatis*) of "physical violence" can be humanly justified. At the next stage we find "organizational compulsion". It is impossible to decide a priori for or against this. It can just as well become the tyrannical power of the authoritarian state as the comprehensively compulsive power of fraternal community. It is nothing if it is dominated by cold calculation, or by tradition and dogma; it is everything if it is borne by "the power of the spirit, by fraternity, by freedom and love." Communists of the stripe of Stoecker have decided in favor of physical violence. They believe that in this world evil can be eradicated by evil. But "every guilty deed is avenged on earth"—and so their activity is condemned to sterility. Social Democrats ranging from Crispien to Dittmann, however, are attached to organizational compulsion, but only that of the organization. In that, too, guilt already resides, a sin against the spirit. Yet for them both roads are still open. To this day, though, we do not detect that breath of love that is needed before any organization gains the right to existence. Otherwise the pathetic drama would not be possible in which Däumig and Dittmann can mutually reproach each other with having social pacifists and Tolstoyans on their side.

If Fell now believes that he can justify the exercise of violence with the argument that the leader may not abandon his troops, he is fundamentally in error. It is as improper for a leader as it is for anyone else (I see no distinction) to commit a lesser evil rather than a greater one; once recognized, evil in any form is to be resisted. Only then can "it never be exercised"! The conscientious revolutionary who not only considers the momentary revolt but has the fate of mankind in mind will never for a moment consider the use of violence. He would always rather die at the hands of the mob than seduce the people by an uprising that leads them astray from their path. He will never be party to the conquest of the socialism of the spirit and of brotherly love by the dictatorship of some authoritarian organization or by violence blinded by rage. He will always be determined in all matters by this highest stage of socialist power, and organizational compulsion will never mean more to him than a necessary instrument.

*Walter Landauer*, Heidelberg.

[32 (Kropotkin); 33 (The "*Abend*"); 34-40 (various book reviews) omitted.]

## 41. PETER KROPOTKIN

*Der Aufbau* (Zurich) II (9) 3/4/21

One of the great leaders of humanity has passed away! an implacable but never embittered fighter for his ideas, an inspired defender of the most sacred human rights, with his always tireless admonition of all the duties of the individual in society. Peter Kropotkin, a Russian who never disowned the best character traits of the Russian popular soul, a man who always held the cause of mankind and of humanity as high as the skies above all nationalism, was surely the most important personality possessed by socialism in our time. On February 8 he died in Dimitrov, near Moscow at the age of 78.

The despotism of Tsarist Russia had made life impossible for him in his homeland. He first visited western Europe for a time in 1872 and, in the circles of the "International Workingmen's Association," in Switzerland, acquired important insights into the revolutionary tasks which he wanted to apply in Russia. His participation in the Tchaikovsky Conspiracy in Petersburg in 1874 led to his arrest and imprisonment in the Peter-Paul Fortress. Two years later, following a daredevil escape, he succeeded in getting away. And so began his stay in the West which lasted over 40 years. From London, the first destination in his flight, he came to Switzerland and there devoted his efforts primarily to the Jura Federation. At that time he was the editor of *Le Révolté*. Expelled from Switzerland, for three years behind prison walls in France, he finally found refuge in England which made it possible for him—despite continued persecution and agitation by the Russian secret police—to do fruitful work. Even in early youth, during his service as an officer in Transbaikal, Kropotkin had been able simultaneously to pursue social activity as well as scientific research, and so to lay the groundwork for his later pioneering studies. His growing awareness of the lunatic conditions prevailing in the social life we lead, made it impossible for him quietly to devote himself to science, while around him hunger and distress abounded. Thus he rejected the offer to become secretary of the Russian Geographical Society so as to be able to dedicate himself to mankind, even though the activities proposed to him substantially corresponded to his talents and inclinations—he had after all already published epoch-making geographical discoveries. In his serious scientific work he had noted that "Science has something of the sublime," yet the consciousness grew within him at once threatening and admonishing that even the researcher and scholar has an egotistical purpose in his own *personal* satisfaction. He was increasingly impelled to pose the question in a radical fashion: "Damnation! you say, if it turns out that abstract science is a luxury, and the practice of medicine is a false illusion; if law is injustice, and technical discovery is a means of oppression; if education, which struggles for wisdom in life, is undermined; if art without revolutionary intention leads only to decadence—what then is left for me to do?" The answer for him could no longer be in doubt: "Either one acts falsely against oneself and says: Let mankind perish, so long as I can flourish! Or one becomes a socialist, and aims at the restructuring of society."

It was these trains of thought that exercised a decisive influence on Kropotkin's life and actions. His early and intensive preoccupation with problems of natural science, especially with biology, did not however remain sterile. It was precisely for this reason, by the brilliant synthesis of his scientific knowledge and socialist convictions, that he was able to unite them in one of his major works *Mutual Aid in the Animal and Human World*. This book is a flaming protest against the nonsense of Social Darwinism, against the foundation of society on Hobbes' postulate of "the war of all against all"; it is a painstaking demonstration of the decisive significance of mutual aid and solidarity in the animal world as well as in human society for all progress; a summons of the individual to the society in which alone he can achieve his highest goals.

This is not the place to go into all of Kropotkin's works. Yet, because of its fundamental importance just for our days, we need urgently to refer to his study on *Agriculture, Industry and Craftsmanship*.

Kropotkin, an extraordinarily acute thinker, was one of those rare people who sense the strength within themselves to *live* their thoughts. As far as at all possible he sought to be his own master in the realm of the mind as well as in all personal affairs: to be a free individual in the service of society, living in accordance with the norms of the loftiest socialism. He took suffering for granted, as a consequence of protest.

"Be yourself!" that is Kropotkin's call to us, as to coming generations. He hated nothing more than every form of authoritarian force. It made no difference from which side or for what purpose force was exercised. Thus it might seem tragic that the aged revolutionary who, after decades of exile, returned to his homeland with high hopes in 1918, found himself again in exile. For his highest principle, that every social grouping, all human society must be based on voluntary acceptance, was turned into its opposite in the Russian revolution!

We mourn in Kropotkin a man who appears once in centuries. But through his work mankind will advance to a massive degree, his seed will germinate in the springtime of human society, and future generations will honor him as the harbinger of moral and spiritual revolution, as the profoundly prophetic student of socialism, as an implacable rebel against the spirit of violence, one whose love extended to all the poor and those bowed down in oppression. It remains our duty to bring his message home to people, to keep his spirit alive!

Walter Landauer.

[42-58 (various reviews; German Settlements articles) omitted.]

## 59. PRODUCTION COOPERATIVE, CONSUMER COOPERATIVE AND SETTLEMENT

*Der Aufbau* (Zurich) II (25) 6/24/21

My primary aim in the previous report on the German settlements was to speak of the perspective from which our most important settlements proceed, and of the practical steps they have taken in order to realize their ideals. Quite consciously, even deliberately, I thereby shifted a major part of the labor of conceptualization to the reader. Anyone who is out of sympathy—and here I particularly mean that sort of interest not too rarely encountered in socialist circles hungry for sensationalism but short on active participation when it comes to new ideas—anyone, then, who is devoid of inner sympathy for the settlement work so rich in sacrifice of which I have attempted to sketch a picture, would have gained just as little profit if I had tried then, or wanted to do so now, to speak in detail of how I would like to see the development of settlements. My objective was rather to place material that would advance the tasks that have been begun in the hands of those who see in the settlement a fundamental factor in the future structure of the community to which we have committed ourselves and in the development of which we must take part. Some have inferred, indeed, from the few words of criticism which I worked in (notably, for example, with reference to the payment of interest[1] that seemed

[1] I would like to add here a supplementary reference to the *Neue Gesellschaft* [New Society] Cooperative for Settlement and the Promotion of Production, Vienna I (Himmelpfortstrasse 9), whose constitution is instructive and worthy of note. No interest, the inalienability of land and the means of production, the limitation on the possibility of acquiring profit to one share only for each member, are the foundations on which they advance their work.

to me particularly damaging for the basis of the settlement, posing a special threat to the likelihood that they will remain faithful to their highest aims. One reader or another may have concluded that I dwelled with particular affection on the Habertsdorf settlement, comparing it, for example, in contrast to the Bergfried settlement which so far is not free of all sorts of capitalist involvements. Well, I do acknowledge that [Habertsdorf] of all the settlements described comes closest to my expectations. But nothing is further from my thoughts than to reject in advance other kinds of settlement on that account. All individuals, like any group of spiritual kin, create and shape ideals for themselves, ideals that are constantly and necessarily modified in the course of world events, become dated and overthrown—but one should always maintain the awareness that they aspire to the fulfillment of one general fundamental idea, desire and hope, to the idea of the loftiest human community. Nothing, therefore, seems to me more inhibiting to the development of settlement questions than meanspiritedness in the judgment of efforts to resolve the common task from different approaches. And, of course, one should never forget that all these settlements are at this time only just beginning to develop their own paths, based on community efforts, and that their laborious progress advances from the founding plan to new experiences, and to corresponding changes in their mode of life and labor. And as long as the basic premises are maintained (about which there will surely be many further discussions in these pages), then the purely formal aspects are not the essence of the settlements. That said, I come now to a few observations that still seem to me to be needed here.

Anyone concerned with settlement problems will quickly encounter the question of the importance the settlements have for the cooperative movement. That estimation will be very varied, depending above all whether the judge attaches greater value to the production or to the consumer cooperative. But for this very reason, however, the judgment will mostly be false, for in the last analysis this dichotomy is irrelevant and erroneous. One can indeed amply discuss the question whether the first step toward a socialist-cooperative community is either the production or the consumer cooperative, but in doing so one should not for a moment forget that whichever one is chosen first, the second step must be for the other. For only both of them in collaboration are viable within a socialist structure, and can overcome capitalism only through mutual aid and complementing each other's strengths. The question of a productive *or* a consumer cooperative does not exist; the question—of course an important one— is simply a matter of conserving energy and time, in deciding which of the two one believes should be first addressed at any given time.

Let us first dwell for a moment on the idea of the cooperative as such. What has first to be said is that cooperation in itself implies nothing whatever about socialism—the word "cooperative" merely denote a *common interest.* That is equally applicable to both the production and the consumer cooperative. We surely prefer to understand by "cooperative" a clearly defined type of common interest, with mutual aid as its primary aim; but we must allow for customary usage so as not to heighten the alarming confusion of words and concepts. Thus I must now direct myself against some observations of Hans Müller ("On the reform of the cooperative movement" in Nos. 10 and 11 of the *Aufbau*). The terms "business" and "production cooperative," and "management" and "consumer cooperative" are not synonymous. But when Müller abstracts from these terms that the business cooperative is necessarily "not anticapitalist," i.e. is capitalist; but that the management cooperative is necessarily "anticapitalist," then one is as false as the other. If one thinks Müller's arguments through to their conclusion, then one comes to the fact that it is a logical inference from his approach that the realization of socialism is not possible because the business cooperatives, and therewith production altogether, *must* operate "in accordance with the principles of an ordinary businessman," i.e. non-socialistically! But it seems all but absurd to me, in view of the innumerable bourgeois capitalist management (consumer) cooperatives to say this kind of cooperative is in principle anticapitalist. No, neither the one nor the other is correct. Let us repeat: the cooperative in itself has nothing whatever to do with socialism. Thus what remains is to distinguish between capitalist and socialist cooperatives. And now we can say: the capitalist production cooperative aims at *increasing the yield* of the individual cooperatives, thus *directly* furthering capitalism, while the capitalist consumer cooperatives, in contrast, *rationalize the consumption* of the cooperators, thus *indirectly* furthering capitalism. Why one should accord one of these some preference I cannot see. In sharpest contrast to both of them stands the socialist cooperative, which draws its strength from sources that in today's customary party socialism are unfortunately all too often diffused. We do not aspire to establish socialist cooperatives in order that people *acquire more or manage less,* but in order to bring both production and consumption into harmony so as to serve the needs of all people. To go further into Müller's observations is impermissible given the scope of this newspaper. But one thing still seems important to me: It is not legislation from without guaranteeing the flourishing of the cooperatives that is most important for the actualization of socialist cooperatives, but rather the autonomous laws borne within each of the cooperators that must be brought to the common task!

If we cast our eyes back, the innermost sense and coherence of principle in the socialist production and consumer cooperatives has become still clearer. I have already underscored that if, in propaganda and appeal one or the other of these is to be given preference, that signifies only which is to be the first step, *with the other to follow as soon as possible.* If it should prove impossible for one to be realized on the heels of the other, then I too would be inclined, in contrast to earlier views, to give priority to the consumer cooperative. But this is solely for practical economic reasons. Both types of cooperative must spring up in the midst of a capitalist environment. Because today it is surely easier to satisfy consumer needs in cooperatives than to establish independent factories of a craft or industrial nature. But even here it will very soon be possible for much to be undertaken, so that socialist production cooperatives should be able to satisfy the common need.

If we now turn back once again to the settlement, we see that there both production and consumer cooperatives are built in from the start. That is what gives them their characteristic quality. The production cooperative, which is undoubtedly the tighter bond more closely defining a community of individuals, is represented by the same people who constitute the consumer cooperative. But it would fundamentally miss the point to locate the essence of the settlement in the expectation that it would overthrow the capitalist economic order overnight. It is unwholesome to abandon oneself to such hopeless illusions in matters of this importance. The settlements will surely face in the future immediately before us—at least in the first years of their existence—a difficult struggle for survival that will be mitigated only by mutual aid. Not even propaganda or legislation will be able to alter this much, though in both respects there is ample work to be done. I see the true sense and import of the settlement rather in their mission as educators for community work and community life. And so too the greater value of the productive cooperative lies in the role of the settlement as the educator for community life through community work. Thus I see the high calling of the settlements in their increasingly training people who develop the socialist cooperatives as a wholly new structure of true humanity. From a community of unabashed self-interest, indistinguishable in its essence from the capitalistic, the socialist community should rise up to become a community of sowing and harvesting, of will and action. Its detachment from the capitalist environment leads the settlement into a socialist community; the League of Cooperative Communities will overthrow capitalism and tear down the walls that separate individuals as well as peoples!

Walter Landauer.

[60 (review) omitted.]

## 61. LENIN, BOLSHEVISM, AND CAPITALISM

*Der Aufbau* (Zurich) II (27) 7/8/21

***Prefatory remark.*** I would like to preface this valuable contribution by Landauer, who draws attention to the latest developments in Bolshevism—perhaps its end!—with the brief observation that I am not unreservedly in agreement with him in the moral evaluation of the Russian Bolshevik leaders even today. As much as I grasp his perspective, and as much as I join with him in rejecting the Bolshevik system as such, yet it seems to me all the more that struggling Russia displays features of heroic greatness even among some of its leaders. We experience often enough the tragedy that it is precisely an originally lofty moral intention that can lead men into the worst of excesses. But, without a doubt, the point of departure for many Bolsheviks is the feeling, springing from well-considered moral sentiments, that tsarist and capitalistic "order" "can no longer be borne" and from the belief that they see the path by which this order can be destroyed, and which leads to the reconstruction of a new community. But in travelling this path they were obliged to suffer the typical fate of the moral fanatic who, obsessed with the particular expression that the idea has formed within him, increasingly loses his way from the idea and its living source of nourishment and more and more in quest of the means forgets and betrays the end. Finally, the Bolsheviks are also to be understood as children of this age, and their mistakes are logically entailed by Russian nihilism and by the Bolshevik version of Marxism. "We are unable to be moderate in anything, to find a middle path, but are always ready to lose ourselves at the utmost extremes and hurl ourselves into the abyss." (Kriskowitch).

Two books have lately appeared, which we shall appraise in greater detail, which have something to contribute to an understanding of contemporary Russia: A. Stolitcher's *Three Months in Soviet Russia*, and Panin's *Tsarist Russia*. Anyone who has read Panin's book will be speechless at the dreadful fate of the Russian people over hundreds of years; it silences almost any criticism of the Bolshevik horror; he will have to satisfy himself with opposing any comparable developments among ourselves.

Max Gerwig.

Innumerable texts have been written for and against Russian Bolshevism from the most contrasting perspectives. Sometimes rejection of particular actions of the Bolsheviks and attacks have been directed against them, sometimes against their entire system and its fundamental premises. But the ultimate question always remained unanswered: how to judge the conduct of the Soviet dictators by ethical norms, to address the question of their good faith. One would have to examine the pros and cons a thousand times over before one could decide whether to concede confidence and trust in his actions to one's opponent. This is what we are obliged to do today in the case of Russia's leaders, as painful as that may be for us in considering the fate of its people. What Lenin wrote recently in a newspaper which just appeared in Petrograd is in so unscrupulous and unvarnished a form, the opposite of what we call socialism, that we cannot conceive that there are anywhere on earth workers who expect anything of him, or of the ideas he represents, or of Russia today. Lenin has proclaimed the bankruptcy of his sytem in clearer words than ever—but by no means what the French call *faillite* [bankruptcy], a failure brought about by unfavorable circumstances, whether internal or external, but the most desperate form of *banqueroute*, a deceitful bankruptcy from sheer dishonesty. That Lenin and his collaborators have elevated "State Socialism" as the highest goal of their efforts we have long known from all their writings and speeches, and it is against this proletarian centralism raised to an ideal to replace that of the monarchy that our substantive battle is and was directed. The cynical acknowledgement that decades of struggle by fire and sword would be required for the attainment even of this state-centralized economy by means of a dictatorship of the *proletariat* would, one must believe, have made difficult to the point of unendurability any justification for taking this path despite this. What do all those have to say now who proclaimed, far removed from the field of battle, that the burden of these unspeakable sufferings by the people of Russia is to be expected and must be demanded in the interest of the final goal and of world revolution; what do they say now that Lenin casually adds that "Shortages and disorganization are so great that we are in no condition to revive State social production on an adequate scale. We must therefore support the development of small industry. That means a rebirth of *petite bourgeoisie* and of capitalism. It would be ridiculous to close one's eyes to this perception." Is there really anyone, on hearing these words, who is not assailed by questions whether it was necessary to take nearly four years to come to recognition of this complete failure to establish a new order of society, and whether it is possible for a normal and ethically thinking person, despite these admissions, to continue to determine the destiny of a people along the same old lines in the name of the *proletariat*. We are obliged to negate both questions! In his further observations Lenin distinguishes four kinds of State Capitalism which "represent a progressive step in comparison with the patriarchal petty-bourgeois element"! They are: concessions to foreign capitalists under governmental control; the cooperatives; the deployment of commission agents to sell State products and to purchase from small producers; and the leasing of State factories and natural resources to capitalist entrepreneurs. It seems to me unnecessary to add even a word of commentary. From now on, anyone who promises himself anything from these people that has even the least connection with the words socialism or communism is either a fool or a knave. How through and through capitalistic is, for example, the concept of the cooperative: "It advances small capitalists into the foreground. The freedom and the rights of the cooperatives signify freedom and rights of capitalism."

And what has Lenin to say about the uncharted, purely agricultural landmasses? Could they perhaps, even if this is impossible for the cities, be turned over to socialism directly? "Yes, this [transition] is conceivable, but only under a single condition: the electrification of the countryside." As often as one reads these words one could only imagine what the population can feel on hearing the announcement of this panacea for their sufferings. And one asks oneself again whether these words, which will remain a curse on Lenin throughout history, can come from a normal and ethically thinking person? For here, too, it is conceded that it will take decades to reach this point! And now we take Lenin's words once again on the conclusion which he has only now drawn from his activities, of which, alas, the Russian people will not soon be able to rid themselves; the conclusion to which dogmatic Marxism is always necessarily impelled: "Capitalism is an evil in comparison with socialism. Capitalism is a good in comparison with medieval conditions, with small industry. To the extent that the realization of a direct transition to socialism turns out to be impossible, capitalism is unavoidable to some degree as a path to small-scale production and barter."

Walter Landauer

[62-70 (various reviews) omitted.]

## 71. THE REALM OF THE ANTICHRIST

*Der Aufbau* (Zürich) II (32), 8/12/21

Gorky recently sent out an appeal to the world on behalf of starving Russia. Undoubtedly with the approval, at the behest of his superiors, who may believe that he would enjoy more trust and success in such a mission than they themselves. And we do not have the slightest doubt that distress and misery afflict the Russian people to an unspeakable and overwhelming degree. At around the same time Mereshkovsky[1] together with Zinaïda Hippius, D. Philossopov and W. Slobin have published a book that is likewise a call for help and rescue for Russia, if of an entirely different kind. One concerted shriek of tortured people, Russians who love their native soil and mankind, who see their own house in blazing flames, and fear that the whole of Europe can catch fire and be consumed by this conflagration. They believe that there is a way between a first tsarist slave Russia and a second "Bolshevik rabble Russia" to a

[1] Dmitri Mereshkovsky: *The Realm of the Antichrist*. Drei Masken, Munich, 1921. 32.50 Marks.

third Russia of free people; they believe in the possibility of this development and seek to promote it—but at the same time they know that this is impossible without the help of the Western nations. And that is why they see no option other than to support the struggle against the Bolsheviks. "In order to topple them, you would sacrifice one or another of them, but we want to sacrifice everything. You can reach an agreement with one or another of them, but we cannot. You want to conclude peace, but we are irreconcilable." The fundamental premise from which this book proceeds is that Europe condemns itself to death if it believes that passive resistance or trade compromises can resolve the Russian question—condemned to death with unqualified certainty.

Profound and, as it seems to me, very significant insights are afforded us into Russian conditions. In judging the character of Lenin, two quotations seem to me very telling in illuminating his thought. Of course, I do not mean by this to say that one can judge a person fully in two sentences, but rather that they fit into the framework of what we know. *"Socialism is calculation"*—with this Lenin displays the quintessence of *his* concept of socialism, and at the same time rejects everything that cannot be encompassed by calculation, and which seems to us the most important. And now the other side: "What are you doing? You are driving Russia into the ground!" someone said to Lenin. "We spit on Russia! Russia is only a basis for us. We don't need Russia, we need world revolution." And this, despite the cynicism, is the truth! Soviet Russia, standing on its own, cannot endure.

"We are all standing, like Ivan Karamazov, between two reptiles, the bourgeois and the Bolshevik. We must choose one or the other of them. But the two reptiles are so similar that one can hardly distinguish them. The Bolshevik one claims to be red, the *bourgeois*—white; both are however bright and spotted; the white reptile has red spots and the red has white ones. And they embrace one another and unite in unnatural concupiscence, either in 'world revolution' or in the cosmic swindle of the concessions that Lenin makes to the Americans; neither can be separated from the other. They lie together encoiled, in love, and snapping at one another. Which will gobble up the other? Neither of them. But from the two reptiles a beast will be born that will gobble us all up. What kind of beast is that? I could name it, but you would not believe me anyway." This is the dismal underlying mood in which Mereshkovsky judges the situation. But it seems to me that Zinaïda Hippius sees the prospects more accurately when she says: "No *qualitative* changes are possible so long as the Bolsheviks are in place, even if they last another ten years; only quantitative changes can be expected ..." That entails still greater misery, the crushing of all spirit, the crushing of all individuality. "Bolshevism is a mental disease, and not only as a defect of rationality, but also of morality—

moral insanity [in English]—no less than an absolute suppression of human personality. In this sense, there are fortunately only very few 'honest' and 'saintly' people among the Bolshevik*s*. But these, however, are the most terrible!"

One might not agree with much that is in the book, as I for example disagree with the judgment of Tolstoy; one could be doubtful whether we have reason or are right to promote "a world-wide Christian league"; but that is a minor issue. The main issue for us is to hear criticism about Soviet Russia from people who wish nothing more earnestly than a free people's Russia. The main objective of this book would be attained if the appeal of Z. Hippius to the workers of the whole world were to be achieved:

"I would like to propose the following to the workers in every land: Each country should elect two plenipotentiaries, two persons whose honesty it trusts (or cannot two absolutely honest people be found in every country?) and these should travel *incognito* (or semi-incognito) to Russia. Besides, honesty, courage and fearlessness are also necessary, because such an undertaking is a heroic deed. I cannot indeed believe that two such self-sacrificial persons cannot be found in every European country. And they will be able to say, following their return (*if* they return) 'to everyone, to everyone, to everyone': Whether there is a revolution in Russia? Whether there is a dictatorship of the *proletariat*? Whether there is a *proletariat*? Whether there is a government of 'workers and peasants'? Whether there is anything at all that resembles a real-life implementation of the principles of 'socialism'? Whether there are soviets, i.e., whether, in the institutions calling themselves soviets, even a trace of the elective principle is to be found?

"The emphatic *No* with which *honest* people, honest socialists, will answer these questions will reveal the fundamental lie in the whole system."

Walter Landauer.

[72-73 (reviews) omitted.]

## 74. ALEX. VON GLEICHEN-RUSSWURM

*Das wahre Gesicht, Weltgeschichte des sozialistischen Gedankens* (The True Face, World History of Socialist Thought). Otto Reichl Press, Darmstadt, cloth 45 Marks.

*Der Aufbau* (Zurich) II (33) 8/19/21

The longing of the masses for "socialism," that is activated in the most varied directions, powerfully excites interest in the history of its ideas. Many component parts have already been furnished us toward a comprehensive history of socialist thought, and Gleichen-Russwurm has enriched us with a particularly valuable one. It is a contribution to the history of socialism, meaning that the book lays no claim to being a complete history of its ideas. Thus, if only to cite two names, I miss Godwin completely, and Bakunin is only fleetingly mentioned. But that was not Gleichen-Russwurm's intention. He aims at depicting individual shapers of socialist thought and showing their connection to its contemporary development. The introduction on the "spirit of historiography" is excellent and lays a foundation for the whole book. On account of the introduction alone the book is an accomplishment and a bitter necessity. That the author is at serious pains to write his history in the spirit of these observations elevates the book well above the level of conventional history. The point of departure is the profound words of the *Talmud*: "A man remains wise so long as he seeks the truth; when he believes he has found it, he is a fool"; this supplies the historian with eternally valid rules for his responsible office that are hardly ever followed. One must first be clear about the fact that there is no such thing as objective historiography. "Any thinker who contemplates events and seeks to record what he knows about them is the seducer of his own soul and is biased. And if only one seriously inquires by whom and to what end history has been and will be written, then it becomes only too clear why it constantly serves the most partisan power interests in the service of persecution. Anecdotes and slogans triumph with a smirk, while the noble visage must remain bitter, rendered silent even after its death." Thus it has been in the history of religion, and thus in political history. Besides the role of religion as the most important instrument of education, its pernicious influence ever increasingly becomes evident. Thus it has come about that "political" in common parlance has come to be equated with cunning and treacherous. And what I have repeatedly affirmed is here elevated to a leading idea: "The political is that which deals with common sense and common weal." No one can be apolitical. Everyone, in every action that in some manner affects his neighbor, is political in the true sense of the word. "It is a fatal stroke when politics and morality are separated from one another."

It is significant throughout the book that its author regards nothing as more destructive to a perspective on history than the materialistic point of view. Gleichen-Russwurm is seeking the truth, and nothing else. And even more important to him, perhaps, than telling people what the truth is, is to stir them up to seek the truth. "Our tragic guilt lies in the fact that we take much too little trouble to attain purity of feeling guided by righteousness and incorruptible truth." For "nowhere in the realm of knowledge have lies such long legs, able to scamper spider-like across long pages of books, weaving webs of forgetfulness and misunderstanding around the noble, around the finest exemplars and greatest deeds" than in history. Beneath this centuries-old web of lies its face is concealed—to make its "true face" visible again should be our most sacred duty.

On this foundation Gleichen-Russwurm erects his history of socialist thought, and in his comments this foundation can support the structure. At the center of his socialist belief is the marvellously succinct utterance of the Carthusian Prior Werner Rolewinck (died 1502): "God and the laborer are the true masters of all that serves the needs of men." The details of the book cannot be gone into here. From Saint-Simon to Jaurès, the great socialists pass before us. From the rejection of materialism mentioned above, there naturally also follows in large measure a rejection of Hegel and of the Marxist school. It seems to me that Gleichen-Russwurm is mostly indebted to Saint-Simon, Proudhon, the pre-Raphaelites and Tolstoy for his opinions. State socialism is strongly rebuffed, and in this rejection of the State itself Gleichen-Russwurm, after all he has said, could have gone a step further. Elaborate state structures seem to him to be dangerous, and he says: "In the interests of progress the State must not acquire increased powers, but be limited; its decisions be subjected to completely free criticism, and it should be easy to improve it." And "in order to make a viable State organization possible, respect for it must be restored." We can go a further step beyond the present State in Gleichen-Russwurm's own words: "So active is this corpse that it can still kill with the poison implicit in it." And in this connection it is no coincidence that Gleichen-Russwurm characterizes as the most original idea of Marx the one that he himself most sharply contradicts in his whole doctrine: "The existence of the State and the existence of slavery are inseparable." Marxism is condemned to sterility if only because it has made slogans of the class war and dictatorship [of the *proletariat*]: since "as long as the world has lasted, whatever has been based on hatred and violence has foundered on hatred and violence."

In our times of party squabbles Gleichen-Russwurm's book has a truly beneficent effect. He is not concerned with this or that theory, but solely with ethical conviction. And despite some differing views I gladly acknowledge that this seems to me the best attempt at a history of socialist thought that we so far possess.

*Walter Landauer.*

## 75. WILHELM MARR

***Anarchie oder Autorität?*** (Anarchy or Authority?) Verlag Hoffmann & Campe, Hamburg, 1852, 12 Marks.
*Erkenntnis und Befreiung* (Vienna) III (38) 8/28-9/3/21

It is nearly 70 years since this book appeared. I do not know whether and to what degree it influenced its German contemporaries at that time. But even if one did not know that the first edition is still available from the publisher to this day, one could conclude with certainty from the mercilessly critical nature of Marr, hostile to all forms of compromise, that his words remained without effect.

The influence of Heine is forcefully traceable on every page: these lines were written with the ironic feeling of one laughing through his tears. There is nothing old-fashioned in saying: "Revolutionaries of all nations! When will you grasp that overthrowing a throne is no help, that it is not the monarchy represented in some individual or other, but authority that is your enemy!" Marr was unquestionably under the influence of Proudhon, even though he never speaks of him. His statements on authority and force are classic; I did not encounter a single sentence that is not of profound importance to us even today. A particular impression was made on me, for example, by the short apodictic sentence: "A republic is honest only for a single moment—*when it promulgates itself.*" Marr belongs to our best, most farseeing and clearest spiritual leaders; not too many copies of his book are still available—anyone who has even a minimal interest in our movement should read it.

*Walter Landauer*

[76(Murder of Erzberger) omitted]

## 77. KARL NÖTZEL

***Einführung zum russischen Roman.*** (Introduction to the Russian Novel.) Musarion Verlag, Munich, 1920. Pbk. 18 Marks.
*Der Aufbau* (Zurich) II (34) 9/2/21, pp. 211-2

Whoever reads this book, which I believe to be among the most significant publications of last year, will be extraordinarily astonished at how much more the book yields than its title implies. Even the subtitle: "Essay on the interpretation of Russian spirituality and of Russian formalism" is only a modest indication of what the author offers. In fact, outside of Nötzel's own earlier writings, no one has so far penetrated so deeply into Russian being and thought as he has here. Based on the Russian novel, a profoundly gripping analysis of the soul of the Russian people and of its poets is furnished; their development and their unique significance explained. The great Russian poet, like his German counterpart, teaches respect for mankind. But the Russian, in doing so, thinks primarily in terms of his capacity for suffering; the German on his mission to create. But an adequate concept of respect for mankind is not yet known to the Russian individual. The Russian's spiritual experience, i.e., of the world outside his own person seen in terms 1) of the purely conceptual aspect is the experience of every sort of abuse of persons by other persons (the yoke of the Tatars, serfdom, despotism); 2) of his own will, the rejection of every form of violence by persons against other persons (this elementary disposition of will constantly finds its expression in the *artel* (collective) and in rural communism as an inclination toward the pan-fraternal); 3) as an emotional experience mediating between understanding and will as his own suffering within the common suffering of all (the Russian is dominated by social compassion: "the shame of not suffering when he sees the community suffering"—Tolstoy). Thus, Russian reality and the Russian soul, both of which are exhaustively treated, together form the elements of the Russian novel. The Russian poet never wholly abandons the opportunity of expressing himself in an entirely personal manner, and in this connection Nötzel correctly says: "we shall have to revise our thinking in order to acknowledge many an aesthetic prejudice. In the course of this we not only learn that tendentiousness as such is not the death of art; and we begin at least to doubt whether in art it is only a matter of "how?", and whether "what?" is not also substantially at issue. It belongs to the deepest essence of the Russian poet that he has consciously grasped that it is his task to lay the preparation for the great reformation of mankind. On the one hand, this preparation consists of presenting to people the reality within which that reformation must become an actuality; and on the other hand enunciate the rousing call to a new outlook that is its foundation and precondition. In the fact that this outlook is not yet present the Russian sees "the unavoidable, personally impersonal social guilt of each individual" which expresses itself in a certain limitation of his self-esteem, and the inner struggle of each individual rages between this and his utter heedlessness—i.e. in the absence of all guilt. This is where the religious problem also enters the Russian novel. Russians are the most extreme subjectivists, even though they take themselves to be particularly objective just on account of the nature of this subjectivism: Russia is "the most powerful human protest raised at the eleventh hour against the mechanization of man, the deprivation of soul in God's creation around and within us." Nötzel thus sees in Russia the spiritual bridge between Europe and Asia, and increasingly it rediscovers its affinity with the Orient, above all with India. But the path of Asia is not that of Russia, which needs the West and its training in historical thought, which Russia lacks, in order to find the correct attitude towards mankind and reality. "A Russia wholly enlightened, a critical Russia, would represent the pinnacle of mankind at this time—no less than a totally and spiritually illuminated Western Europe would. India would transmit to both an Indian reality and learn from it the ultimate refulgence of the purely spiritual."

I am well aware, and it is necessary to stress this, that I have sketched only a meagre silhouette of the dazzling structure of thought in Nötzel's book. Since what he does in this book is to summarize the conclusions of many great works, the book is particularly suited to give us a picture of his interpretation of the Russian nature. And it seems to me more important and fruitful to penetrate the essence of the Russian people and its spiritual heroes than that of the Bolsheviks.

[Unsigned][1]

[78-85 (assorted reviews) omitted.]

[1] We have included this among Landauer's writings, although unsigned, for a number of reasons, none conclusive, to be sure. He has so far been the only contributor to a column *Gute Bücher* in *Aufbau,* which he apparently started, though later on other reviewers were added. He was interested in and admired Nötzel's work, as we see from other references to him in this collection. The style is Landauer's, notably the reference to the Bolsheviks at the end.

## 86. SOCIALIST LITERATURE

*Der Aufbau* (Zurich) III (8) 2/24/22

Books born in a socialist spirit and which at the same time possess high literary value are unfortunately still not too common in our time. All the more important is it to draw attention to them. At the head of the line among our contemporaries stands the Dane, Martin Andersen Nexö. One would first mention his novel *Stine, the Child of Man* (Verlag Albert Langen, Munich), of which four parts have so far appeared. It unfolds for us the story of the development of a girl whose path leads through all the filth and unspeakable misery that are the most faithful companions of poverty. With extraordinary dedication to the truth, sparing us nothing of all her sufferings, the portrait of Stine's character is drawn, but at the same time we see the psychological and social conditions that determine her actions. At every moment we grasp how, in the struggle of character traits imposed by religion and science, individuals with unfavorable inheritance and social disadvantages shape their lives. We see this struggle in Stine who, even under the most oppressive misery, remains a human being to whom all our love and respect are attached. A joyful purity and courageous affirmation of life give Stine the strength to come through all the entanglements and to stand firm even in the face of crime. The most varied images pass before us—beginning with the minor sufferings of the illegitimate child to her job as a nursemaid in an "angel factory" and the vicissitudes of her experience as a servant-girl. Here she once found satisfaction, happiness and love—and many aspects of her loving and warm nature only then emerge. At this pinnacle of the depiction he shows the great importance for human action of external conditions. But we repeatedly see also—and that is significant enough—that it is neither the social and other external conditions, nor just the innate inner preconditions that alone determine human existence, but rather a synthesis of both. "Stine, child of man" represents in total the efforts of a personality endowed with high moral gifts to achieve her own identity despite all the impediments.

Even more important for us, however, is Andersen Nexö's *Pelle the Conqueror* (Insel-Verlag, Leipzig). At its core there stands once more the development of a man who elevates himself from the dismal existence of a shepherd-boy to the victorious leader of a socialist movement. It would carry us too far afield to go into details of the personalities described and of their destinies. But it is important here to emphasize one thing. The work is ideal in its synthesis of art and propaganda in its best sense. The fundamental socialist thinking is in complete harmony with the people represented, and of their nature and character. None of them is a mere framework for the depiction of any other; rather, the whole narrative constitutes a solid fabric in the unified representation of living people.

Let us now briefly turn to the socialist thoughts that lend the book inestimable value for us. Even before Pelle joins the socialist movement he is told: "Nothing good is ever achieved with the aid of evil," and he does not suspect how significant these words will one day become for him. In the struggle against industrial exploiters he first lends support for uniting in craft leagues (trade unions). A successful boycott, which he organizes, leads to the establishment of the socialist party, and finally to the *Internationale*. A vigorous struggle against the capitalists, thought by his supporters to be a decisive battle, does indeed bring some concessions to the workers, but these are purchased at too dear a price to carry the movement forward to still higher goals. The forces of reaction understand well how, with the aid of compliant judges, to place Pelle behind bars for years. In the meantime the party falls ever deeper into that stagnation typical of our Marxist parties. Strike initiatives, disputes among the unions, parliamentary procedures, all increasingly become ends in themselves. After his release, Pelle finds all his connections in the movement broken off, and now there slowly begins to germinate in him the thought of a completely new vision of the world. He is compelled to think of his former activity as a failure—a failure of the workers themselves. "What they lacked was responsibility. The idea that sustained the movement had apparently passed them by without a trace!" "God knows, we obeyed the law, we sweated out parliamentary procedures, and sought nothing more than to win votes—and then what? Could anyone at all be in doubt that poor people were in a majority—in an oppressive majority? What kind of nonsense was it that the majority should be overcome? No, those in possession of power would know how to arrange matters that they kept it—and you could then win as many stupid votes as you liked!" Pelle began now to understand how right those warning voices had been who had earlier said to him: "What good does it do me if I get people to respect me by beating them? That, damn it, is a matter of indifference! But if they perhaps respect me for myself?—if they say of their own volition: Look, there goes a man, formed in God's image, one who thinks and feels in his inmost self just as I do!" And so Pelle now begins exactly where he had begun before—with the craft league. But this time he knows that the answer to the socialist question must be sought in an entirely different direction. Not in the contests between political parties, not by violence and dictatorship can a new society be constructed, but only by the inmost unity of people of like beliefs, united in trade unions based on mutual aid. "A new time must first come in which everything that is necessary for all to participate—goodness of heart and solidarity—comes to the fore. That great alliance itself, in the formation of which he now participated, could now at last be seen in the right perspective. It had been the antithesis of "One against all"— it had been based on a law of mutuality. The path and goal of this new era can only be a trade-union socialism which (in deliberate opposition to the Marxist parties) did not exacerbate the class struggle, but gradually overcame it; one that did not base the alliance of people on authority and violence—even those of some socialist dogma—but on goodness and knowledge; and which therefore no longer accepts at all what we today call the State. Pelle begins by realizing this in his own handicraft, by founding a union of shoemakers. From this point on "a world-wide peaceful revolution on a grand scale" should come about. "His trade-union activity should become the point of departure for a world war between labor and capital!" The actuality progresses rapidly forward. Neighboring trades are first incorporated. At the same time a cooperative settlement of workers on the land is started to liberate them from the city, because it exercises a leaden power over their spirits, drawing poor people to it by the false promise of happiness. And once the city has them in its clutches, it holds them as by a demonic spell. The poisoned air was the opium in which the most miserable poor dreamed of happiness, and once they had acquired a taste for it they were no longer in a condition to leave it for the plain light of day."

We see how everywhere from small beginnings a great building swiftly rises. The trade unions, indeed, are more or less vehemently rejected by our socialist parties (all the more energetically the more socialist they take themselves to be; i.e., the further left they stand in the political struggles) and indeed, independent of dogmatic party lines, because they believe this path will take much too long. We know that the opposite is true; we know that this route is for the present the only possible one, and that is what is shown in this book. How simple and self-evident is the whole course of thinking by which this new community is to be built up in agriculture, handicrafts, and industry! "Like a good home filled with love in which no one undermines another, that is how the world should be—the whole idea is no more complicated than that." Pelle's supreme principle is: "We who are ourselves little want to show how the little people maintain and validate themselves through the principle of goodness. We intend no evil to anyone and that is why goodness is on our side." From this unshakeable belief springs the strength to lead the idea of communal socialism to victory. We find in this work of Andersen Nexö the sincere thoughts that we have tried in these pages to illustrate in all their significance, and we find them here interwoven in a wonderful poem.

Walter Landauer.

## 87. STATE INHERITANCE LAW AND THE COOPERATIVE MOVEMENT

*Der Aufbau* (Zurich) III (19) 5/12/22

Most readers of *Aufbau* will be familiar with the ideas of the *International State Inheritance Law Propaganda* in Basel. It will, therefore, suffice to lay out here the principles of this program, and for the rest to refer to the literature [1]. In the draft Bill that has been filed it states: The State is the universal legatee of its citizens. The State considers spouses, as to their rights of inheritance, as a unit. Only when *both* spouses are deceased does the State's right of inheritance become valid. Items in an estate that have the character of sentimental value may be privately inherited, provided that the gain in the heir's legacy does not exceed the sum of ___. Gifts during the lifetime of the deceased are treated in the same way as legacies. Excepted from this are objects of daily usage for the personal use of the recipient, provided that their total value per month does not exceed ____. Evasion of this law and offenses against it will be treated as embezzlement of State property and punishable by _____. *As an interim measure, a limited right of inheritance by descendants and ascendants will be considered* (see *Aufbau*, No. 14). Concurrently with this Inheritance Law is a law providing for the establishment of a State Estate Bank (SEB) and to assure the binding payment of checks. The SEB is responsible for all estates presently existing and arising in the future within the country for the totality of everyone residing in the territory of the State, and will administer and evaluate all assets falling to it under the right of inheritance. A checking account will be opened at the appropriate branch of the SEB for all competent inhabitants as well as for all legal persons. The transfer of property, to the extent that it exceeds an upper value established within narrow limits, can then take place only without the use of cash between accounts at the SEB.

These are the principles of the two laws by which unlimited credit is to be supplied to the State, and is to enable a general socialist planned economy to be established. The SEB is able, under these conditions, to assure interest-free credit in any amount, whereby any form of capitalist financing is of course brought to an end; from a purely *economic* viewpoint the State right of inheritance and the obligatory use of checks (especially on account of the associated virtual abolition of money!) must seem a suitable weapon for the definitive elimination of capitalism.

If we wish now to clarify our position on the theory of the State right of inheritance, the first issue—as always—is to establish whether, to what degree

[1] In particular: Paul Tissot, *Die Verwirklichung der sozialen Demokratie mittelst Staatserbrecht und Scheckobligatorium* [The Implementation of Social Democracy by means of State Inheritance Law and the Obligatory Use of Checks], Friede durch Recht Press, Ludwigsburg.

and in what form, we can harmonize the basic principles of the State right of inheritance with our socialist objectives; and whether this has significance for our aims.

All previous inheritance taxes had the sole purpose of making the State a participant in legacies, but otherwise they generally gave a free hand to the private initiative of capitalism. Like all other taxes, inheritance taxes served solely to generate revenue for the State. The State right of inheritance, in contrast, *aims at* and *mandates* a complete overturn of all economic and social conditions. It does not pursue the aim of opening up new sources of revenue for the old state, but is to be the lever by which the capitalist economy is displaced and a socialist one is established in its stead. Therefore, in the case of the State right of inheritance, it is less an issue of the "right of inheritance" as such; rather that by *total* inheritance the acquisition of property is impeded, so that gradually an automatic socialization of economic life is to be achieved.

It now seems to us high time to deal with two questions that must emerge from the foregoing. 1. Is socialization solely achieved in terms of economic activity—which as we have seen is to be brought about by the State right of inheritance? And 2. Upon introduction of the State right of inheritance, who is to be the recipient of the estates falling due? We shall soon see that in the end both questions lead to the same result.

Just to pose the first question entails negating it, even though all too many socialists (notably the party secretaries and union leaders) are unfortunately unclear that "socializing the economy" and "socialism" are by no means synonymous concepts. The realization of socialism must always be pursued simultaneously along two parallel paths. If one of these is favored at the expense of the other, then disappointments and reverses are unavoidable. Most doctrines and programs, especially the dominant ones of the socialist movement, have thus far foundered on this onesidedness. We are referring to the path of materialism or of idealism, or as we would say here the socialization of the *economy* and the socialization of *people.* In those times when economics played no significant role in national life, in which industrialism and pauperism were unknown, or known only to a modest degree, the pioneers of socialism were almost exclusively concerned with the socialization of people. We have in mind Christian teachings, the basic idea of the craft and guild movement. Among the so-called "Utopians" (Owen, Fourier, etc.), who in their deepest being were pure idealists, the reaction begins against the capitalism that was gradually emerging, to effect what was in part a very remarkable synthesis with materialistic thought. The growing material distress of the working masses, which knew no limits, swept all idealism to one side, and led to the slogan: socialization of the economy as the cure for material misery. All these stirrings and sentiments of the *proletariat* were approved by Marxist doctrine and its materialistic philosophy of history; and on an ostensibly "scientific" basis they proclaimed that the evolution of mankind would proceed automatically from capitalism to socialism under the influence of the economic conditions prevailing at any time. The harsh lessons of recent years (especially in Russia) and the bitter distress that still faces us, will compel more and more people, even in the Marxist parties, to recognize what individual socialists outside the parties (and therefore without influence) have long seen: that all attempts to socialize the economy will sooner or later fail until people become socialist in their utmost beliefs, thoughts and actions; until their socialism is effective in action and life (within their families and sphere of influence) rather than expressed in theory and propaganda. That this is not the case today hardly anyone would dare to deny. We shall therefore have to return to the point that the socialization of people must be placed in the foreground of our efforts. Since individual egotism is not impeded by political and economic measures directed against capitalist potential for acquisition; and capitalist thinking is not destroyed along with capitalism and its mode of life; then we should be completely clear that even the introduction of the State right of inheritance would alter only the economic conditions, but not the people. To be sure, there can be no doubt whatever that economic conditions have an immeasurable influence on people's actions (as, for example, virtually all crime, even a large proportion of that caused by psychopathology, is caused by social conditions), but one should not believe that perfecting the economy will of itself be able to produce perfect people. For that what is needed is a broad horizon in the thought of the individual as in the inmost beliefs of the masses. Even the State right of inheritance would become and remain a dead letter if, instead of being a lever (means) in the hands of conscientious socialists, it seeks to become the cause itself of a social revolution and imagines one consequence of its effect to be the conversion of people into socialists. The conference in Heidelberg has shown clearly enough that it is precisely in Germany that the danger threatens, that the idea of the State right of inheritance is to be assimilated into the socialization platform of the trade unions or political parties as their finance program. I speak of a danger because thereby the path of centralism rather than of federalism or cooperatives is embarked on, because in that case it is in fact a "State" right of inheritance, rather than that of a community or people's right that is to be introduced. The proposals for the State right of inheritance could certainly play a very important role, but in itself could lead to every outcome from the violence of Russian communism to the comradely socialist community life that is our ideal. Only when it has become clear to more people that the most important requirement is the socialization of men, then things will proceed more rapidly along our path, and only then will the State right of inheritance proposals (along with those in the closely related Wissel planned economy) be able to fulfill its important task. Let us now ask who would be regarded as "the citizens' universal legatee" under our socialist program; it should be clear from the foregoing that we reject the present State, as we do any centralized State, for this role. In Tissot's book he conveys in many passages that even he foresees the fearful consequences of such a concentration of power and property in the State, and seeks to limit this by craft and trade cooperatives. Following this idea to its logical conclusion brings us back to our cooperative socialism in which the individual communities and cooperatives in the city and on the land would be the heirs of their members. The polity, as the cooperative of cooperatives, as the federation of the communities, would then enjoy exactly the same measure of unlimited credit as if it were the direct heir of its citizens, since it is charged solely by the individual consumer, production, and property (including land) cooperatives to protect the interests of the totality (cultural, educational, international issues).

With this needed skepticism, without constantly falling prey to dogmas or surrendering to distorted and sterile illusions; with the needed idealism without which no moral action is possible, we thus come to the following conclusion: We believe that, in the interests of socialism, the State right of inheritance must be unconditionally rejected if, as a financial measure, it would lead to the transfer of a great part of the property and, in particular, of the initiative, of enterprises into the hands of the present-day State. We believe, however, that within the framework of cooperative socialism it would play an extremely important and beneficial role. This cooperative collaboration can be initiated by enterprising people at once; when the passive resistance so far demonstrated by the socialist Marxist parties has disappeared, it could take on extraordinary dimensions in a short time; then the obvious advantages of this voluntary union would permit its promulgation among citizens, civil servants, and farmers; and soon allow the implementation of its proposals through the program of the International State Right of Inheritance Propaganda.

*Walter Landauer, Heidelberg*

## 88. [93] LABOR COOPERATIVES[1]

*Der Aufbau* (Zurich) III (34) 8/25/22
*Die sozialistische Genossenschaft* (Greiz) III (6) 3/21/23

The cooperative movement has already been described in these pages from the most varied perspectives, and has been evaluated for its significance for socialism. Thus far, to the best of my knowledge, there has never been a

1. Like its companion piece on the Tenant Cooperative (#91-92) this was reprinted as indicated above. The text is identical except for the omission of the first two sentences, the word "however," in the third sentence, and the date of Preyer's book in the footnote. [Ed.]

discussion of the "labor cooperative," a form of cooperative that has undergone, in Italy above all, a typical development in conjunction with the agricultural tenant cooperative. In the emergence of new societal institutions from individual cooperative communities, however, the labor cooperative which appears alongside and among the consumer and productive cooperatives is of particular importance as a social linkage. I shall therefore attempt here briefly to give an account of the most important principles of the Italian labor cooperatives (*cooperativa di lavoro*),[1] and thereby hope to stimulate further interest in them in discussions of cooperative issues.

The labor cooperatives in Italy arose some forty years ago in Emilia province, where they are still principally in existence. They came about because of overpopulation and the resulting extended and chronic unemployment, and represent a further development of the trade union organizations first initiated to relieve it. One can readily find many connections to the age-old labor communities in very different countries—something like the Russian *artels*. The immediate aims of the new movement were: raising wages, the rotating of workers in a specific cycle so that each received the same amount of work as far as possible, and a reduction in the hours worked. The labor cooperatives—composed mainly of unskilled rural workers—carried out public and communal improvement projects (like road works, redirection of rivers, etc.) to the exclusion of private enterprises. The assignment of the work mostly followed a bidding process. But they often also took over agricultural operations and so also simultaneously became tenant cooperatives, about which I mean to report shortly in another essay.

If we for the moment ignore these transitions to the tenant cooperative and thence to the production cooperative, and consider the relationship of the purely labor cooperatives to the production and consumer cooperatives, then the following needs to be said. In accordance with their general character, the labor cooperatives are naturally closely related to the production cooperatives, but differ from them in principle in that the members offer only their *labor* in fulfillment of the common objectives of the work, while the production cooperative also supplies the necessary raw materials to be processed, and in addition is concerned with the profit-making sale of the finished products that it has manufactured. For the labor cooperative, the main advantage of this is that they can proceed with a minimum of business overhead. The leadership of these businesses is mostly shared by an administrative council, an executive council, and a technical committee. The administrative council, elected solely by cooperative workers, determines the acceptance of projects

[1] In reliance on the excellent book: W.D. Preyer, *Die Arbeits und Pachtgenossenschaften Italiens* [The Labor and Tenant Cooperatives of Italy], Gustav Fischer Press, Jena 1913.

on the recommendation of the technical committee, the admission of new members (which requires the guarantee by at least two members of the reliability of the applicant), the distribution of net profits when accounts are settled, and the authorization of loans, etc. When a project is accepted, the administrative council determines the amount of total expenditures for its execution (between 15 and 30% of the over-all price), and distributes the remaining sum among the workers in such a manner that the weekly payments correspond exactly to the scale of the local labor department, and whatever is left over is paid to the workers when the work is completed.

In addition to cooperative workers, other persons can be elected to the executive council. Election is for a year (two years in the case of the administrative council), and can be renewed. Its primary activity is to oversee that the cooperative's statutes and decisions are strictly complied with. The technical committee, chiefly comprised of engineers and legal experts, is charged to evaluate and report on projects proposed to be undertaken by the cooperative. In all important respects this is the administrative structure and function of the labor cooperative.—Beyond this there are also connections to consumer cooperatives which, however, are markedly in the background compared to labor projects; most labor cooperatives are, indeed, also consumer cooperatives where the accounting is either separate or jointly conducted.

Most labor cooperatives are of an emphatically democratic or socialist character. Only a few, among them "the best organized and most successful, however, expressly declare themselves opposed to political activity."

In practice the Italian labor cooperatives have attained extraordinarily great success, also recognized by the government by the award of numerous contracts. For the workers an increase in wages and a reduction of working hours are the chief benefits. In this an oddity of the national economy is displayed: "the labor cooperatives, as masters of the labor market, demand higher wages the greater the number of unemployed, thus of those seeking work. Their aim in securing higher wages is to compensate for the long periods of enforced idleness." That is, of course, possible only within specific limits. Under the pressure of unemployment, working hours have been reduced to an extraordinary degree—down to a daily rotation of two hours at an hourly rate of 1.75 Lire (prewar conditions!); under labor pacts a maximum of six hours is never surpassed on public works. The time required for the cooperatives to complete their projects, however, is often less than that taken by private enterprises because they have many more workers at their disposition. "A particular advantage of the cooperatives, it is stressed by officials, is that they less have recourse to the courts than do the [private] enterprises; the claims brought before the courts by them are better founded as a rule."

A. Wilhelm reports (in his fine book: *Wirtschaftsdemokratie der Zukunft* [Economic Democracy of the Future], Anzengruber Press, Leipzig 1919) about the labor cooperatives: "Today virtually all public works throughout Italy are by preference awarded to the labor cooperatives for completion. Considered from the perspective of a people's economy that is the best form, and most promising of success, of "socialization of labor" in the best sense of the word, an idea of which the theoretical super-Socialists do not have the faintest clue, and which all the proposals encountered so far for socialization do not even approach in their practical significance."

To the overview of gross sales by the consumer cooperatives, of production to meet individual needs by the production cooperatives, we must add the labor and production to meet the requirements and necessities of all that is performed by the labor cooperatives. In this connection it is of notable significance, unfortunately insufficiently acknowledged, that this collective work is performed on the joint responsibility of all the workers for the efficient and effective execution of the work.

*Walter Landauer.*

## 89. A SOCIALIST PROGRAM

*Süddeutsche Sonntags-Zeitung* (Heilbronn) III (34) 8/20/22

There is something strange about the word "program." It prompts so many undertones of useless talk, avoidance of the facts, and of illusions remote from reality—for some only slightly discouraging, for others powerful enough to frighten them away. Only those very few for whom socialist "action" is a matter of bitter seriousness are inclined to interest themselves in *Utopias.* That the great majority of socialists have not sought to work for any Utopia that they could make their own is surely one of the worst sins of omission of our time which is so accustomed to omission. For in this is rooted the thrust of the materialistic Marxist doctrine most destructive to socialist aspiration: the dogma concerning the development of mankind unalterably determined by economic conditions. And while the theoreticians of Marxism quarrel with each other whether the capitalist economy in Russia was "mature" enough yet for the transition to socialism, while on the other hand in Germany it would have been sufficiently mature, the awareness arises here and there that most people today lack the *inner* preconditions that are necessary for the realization of a socialist society.

There surely dwells in the great majority of workers, in the members of the various socialist parties, an honest desire for socialism, but its momentum is inhibited by the lack of even a somewhat clear idea; what it is supposed to do. This situation increasingly compels individuals within the parties to the inference that if there is to be an effective transformation of the economic structure, *a new program* and among the people the will to execute it are needed. Among those who did not know it already many have now reached the sobering conclusion how little bearing on the realities a formulation like the Erfurt program has. But there is a program able powerfully to ignite the moral forces in the socialist movement, one that clearly expounds its principles, means and ends: *A Socialist Program* by Max Gerber, Jean Matthieu, Clara and Leonhard Ragaz, Dora Staudinger; Publisher W. Trösch, Olten.

To be sure, the acceptance of this program entails the rejection of many erroneous traditions. We are confident that, sooner or later, the moment will come for a breach with materialistic Marxist dogmas, but we trust that this crisis, which must lead to the restoration to health of the socialist movement, will be eased by the book of these Swiss authors, and its inception hastened. To that end it is necessary—despite its appalling price by German standards—that we have access to it; above all, it should be the obligation of all public and party libraries to make the book available to our people. Although the authors of the program have independently written various parts of it, yet because it has emerged from joint discussions, its inner unity is fully preserved, something assured indeed by the fundamental agreement by the authors in all essentials.

No new party—in opposition rather to all parties; no new dogmas—rather than the old persistent stimulation of socialist currents of thought, the education of independent individuals; these are among the most important principles of the program. "For we see the notable value of our program just in the fact that we mean to contribute to the emergence of this free outlook. We do not thereby misjudge the dangers at all that inhere in the proposal of any program. Nothing is further removed from our purpose than to specify the forms we believe socialism must take for its realization to seem possible and desirable to us. On the contrary, we seek to liberate it from just such dogmatic forms and doctrinaire constraints, and so help to bring it out into the open. We would like to show that it is a current that washes away all these various modes of realization, that strives toward [new] horizons, but still one that [points to the path] where we in service to the future, have our task to perform.[1] So far

[1] A fold in the crumbling newsprint has eliminated the words translated in brackets. The German text reads, with our attempt to reconstruct it, as follows:

> *... wir möchten zeigen, dass er ein Strom ist, der alle diese einzelnen Verwirklichungsformen wieder auflösend, immer [neuen] Horizonten entgegenstrebt, und doch gleichzeitig auf den [Weg erweisen], wo wir, der Zukunft dienend, unser Werk zu tun haben.*

as in our power lies, we aim at stimulating new and free socialist thought and action, but not to create some new restriction on it; we want to release a current, not dam it up."

Marxism has been able to draw virtually the entire workers' movement into its ambit solely by sheathing itself in the robes of infallible science and representing its beliefs apodictically as ultimate knowledge. The modern worker is mistrustful of all 'beliefs.' He looks for *certainty* in his socialism. And that it seems to him only *science* can give. But the opposite is correct: *Nothing is more uncertain than scientific theories, nothing is more certain than the eternal facts of the moral world, and nothing therefore is more triumphant than the belief in them.* That is the rock on which socialism must and should build." Thus it is only on the certainty of eternal moral values that this socialist program is founded. The self-evident precondition for the ultimate objective of this kind of socialism is, of course, "the fullest possible dissolution of the undifferentiated masses into independent individuals, thus total moral autonomy." But this socialist program does not for that reason collapse into the extreme of a onesided idealism, because it is fully aware how far material conditions have to be taken into account. "Spirit *has need* of matter. The whole sense of socialism is not that it seeks to cultivate an abstract spirituality remote from the world, but that it wants to encompass the *whole* world, and that means: to control material conditions for the spiritual."

In the chapters on the meaning and history of socialism, on the socialist movement and socialist parties, a number of admirable observations are to be found on the development and essence of socialist thought and of the socialist movement which, bit by bit, cohere into an over-all picture of a new world. It must be a fundamental principle in the battle for socialism to employ only those weapons consistent with its own objectives, i.e. to *renounce all forms of brute force,* and "the chief danger it faces is that it can be misled into taking the line of least resistance." The integrating role of the *proletariat* in the socialist movement and the complex necessity of *class warfare* are of course recognized, but there can be no doubt that it may never be the aim of an honest socialism to elevate *one* class to domination, but rather that the class war must be overcome by the elimination of all intellectual prejudices and lines of economic separation. The task of the workers' movement, therefore, must not be some particularist "class tactics," but has the duty "to advance the general social and cultural objectives of socialism to the fore; instead of hallowed egoism *[sacro egoismo]* concern for the whole must predominate, the common interest instead of class interest." The weapons of this form of socialism are: propagandizing the *idea; action* (cooperative movement and educational system) and the *strike.* "These three weapons which are those of the spirit and of justice, and not those of violence, are completely sufficient to carry socialism to victory. Heaven and earth in any case cry out for them. Any others would do us harm."

We can only sketch the details of the program here in a few words. Since the socialist economy is to bring about "the dissolution of the contradiction between individual and society," it must maintain a balance between anarchy (as defined by Stirner) and coercive organization; it is "cooperative maintenance through cooperative labor." From this follows the *rejection of the contemporary State* as well as of all forms of *State socialism*. "In any case, once socialism has been realized and been embodied in the economy, it will build its own political body, its own 'State'; but this will have nothing in common with what is called a State today. Socialism is a totally opposite principle. It is rooted in mankind itself and not in institutions, therefore man must grow into it, must emerge as a spiritual force, and cannot be ruled by decree." The new society will be underpinned by the principle of *federalism*; the family and the community are its units. The ultimate form in which the economic body is articulated would be a system of *councils*—a form hitherto never achieved, to be sure—that depends equally upon socialism and democracy, upon the federal cooperation of all economic units. Cooperatives for consumers, for production, and for property (in particular for land and housing) are to replace the present economic forms, whose economic interests will be represented within the several professions by "economic councils." We can begin at any time to actualize such cooperatives, which would be entirely different from what passes under this name at the present time.

We have seen that this socialist program rests on two basic principles: on the one hand the socialization of the economy by the cooperatives and, on the other, the socialization of the society by the creation of free and self-confident individuals. And since, unfortunately, for the effective transformation of contemporary conditions it is lack of maturity in individuals rather than lack of maturity in the economy, nothing is needed more than new forms of *education*. "It is upon this that not only spiritual renewal depends, but also the transformation of all forms of life; it represents the goal that socialism must pursue if it is not to capitulate." What is briefly sketched here is fully expounded in the forthright work by L. Ragaz (*The Pedagogical Revolution*, published by W. Trösch, Olten) on education and is among the most significant achievements of our time.

The purpose of these observations is only to indicate the principles of this new socialist program, one which in essence, while rejecting a political process, strives to found socialism on freedom, community and mutual aid. For that reason we do not need here to go into the transitional program, the agrarian program or the program for a women's movement.

"As socialists we also aspire to a world in which what otherwise disintegrates into disputing theories and parties is unified into a higher synthesis which thereby loses its fanatical onesideness. We believe in the day when this wider horizon opens up in people's minds and socialism emerges from the constraints of party dogma into the grand freedom of a vital principle and mighty belief."

"Socialism has now been placed on the agenda of history. That is the new order of business. Any tactics not relevant to it are out of order. What is then the novelty that this order of business demands? It can be briefly formulated thus: What is now important is that the new world aimed at by socialism should be elaborated, theoretically and practically, with the utmost possible purity, splendor, and clarity."

Together with Nötzel's *Introduction to Socialism without Dogma* (Musarion Publishers, Munich), this "Socialist Program" is surely the first book since the days of the November revolution to fulfill this demand in the broadest sense. Because it is original in its development from the foundations up in advancing new ideas in the realm of socialism, and thus proceeds in a *revolutionary spirit,* we hope that its effect on the socialist movement will be *revolutionary.*

*Walter Landauer*

## 90. RUSSIA'S LEADING SPIRIT

*Der Aufbau* (Zurich) III (42) 10/20/22

It could in no way be a fortuitous phenomenon that interest in the writings of the present leaders of Soviet Russia has significantly declined, and instead that more and more books appear that are representative of the specifically Russian spirit that is particularly important to us, but which is suppressed by the present-day system of government. For there can be no doubt at all that the Hegelian-Marxist state socialism of the Russia of today has inherited little of the ideas of the great Russians of the past century. Whether it is Tolstoy or Dostoievsky, Kropotkin or Bakunin, they are dead now and will be fruitful only in future times. A series of new publications could contribute to smoothing their path in Germany, and therefore deserve special attention. For the first time in German there has just appeared the youthful work of Peter Kropotkin, *A Rebel's Words* (Erkenntnis und Befreiung Publishers, Vienna-Klosterneuburg). Thereby an old debt is repaid. In it Kropotkin lays down the general principles of the views which he has explored in depth and detail in his other works. We find many profound insights into the nature of eternal events, for example, when he says about the revolution: "Revolution breaks out in the history of peoples long before the possibility of there being a general understanding of it, and those who have a clear idea of what, immediately after it has succeeded, is to be done at such a time are unfortunately in a small minority. The great masses usually have only a general notion about the aims they seek to realize without much confidence in the path that leads to it. The practical solution will only then be found and assume clear form when the overthrow has already begun; the outcome of the revolution can only be that of the people acting independently, or it is nothing, for the intellects of a few individuals are incapable of finding solutions that can be born only of the life of the people." Kropotkin's discernment breaks new ground finally in party circles also: "It is not the circumstance that the people were in possession of a comprehensive theory that stood in the way of their acting—no, they had too little theory, or none at all."

It is equally welcome that at last Michael Bakunin's *Collected Works* (Syndicalist Press, Berlin, Vol. 1) is accessible to German readers, for incredibly false accounts of his views have been disseminated. One could surely hear often enough that it was in the first place Bakunin and not Marx whose aims were being expressed in contemporary Russia. It is only necessary to read the book *The Knoutogermanic Empire and the Social Revolution* to see how irrational this assertion is, given that he sums up his views in one passage, for example: "In this system there is really no power. Power resides in the commonalty, and is the true expression of the freedom of each and every one, the faithful and solemn actualization of the will of all; each man obeys only because the leader orders him to do only what he himself wishes." Or in another place: "Man's freedom subsists solely herein, that he obeys the laws of nature because they come to him externally, from some other will, whether divine or human, and are collectively or individually imposed ... Thus there should be no legislation imposed externally, and no authority, each is inseparable from the other and leads to the enslavement of society and makes an imbecile of the lawgiver himself." And then Bakunin develops in *God and the State* an acutely thought-through philosophical perspective, in many respects based on Spinoza, culminating in the concept of "universal causality." This brilliant work deserves the closest attention.

Maxim Gorky: *The Destruction of Personality* (Rudolf Kaemmerer Press, Dresden, paper 28 Marks; linen 35 Marks). This collection of essays from the years 1905-1916 will surely excite lively interest—despite the abhorrent polemic by Mereshkovsky. We do not know for certain whether and to what degree Gorky's views have since changed—as we encounter him here, he surely has important things to tell us. His words are an accusation against Marxist Russia no less than against tsarist Russia: "The State kills the human so that his brute strength magnifies its power; it fights against reason which is always opposed to violence. The welfare of a country rests on the freedom of its people; only their strength can overcome the dark power of the State." This is consistent with the principal thrust of the book which defends freedom and the opportunity for individual development of the person.

F.M. Dostoievsky: *Diary of a Writer* (Musarion Press, Munich, Vol. 1; paper 30 Marks; linen 36 Marks). This volume includes essays published by Dostoievsky in 1873 in the journal he edited *Grashdanin* (The Citizen). We find much and multitudinous, some of it worthless today, but also some of very great significance. We know indeed that the politician Dostoievsky is a reactionary, even if one often has the impression that he has to force himself into these views. And it strikes me as extraordinarily characteristic of the reactionary opinions of the older Dostoievsky that despite everything his penetrating intellect always emerges triumphant. Just one example: "The French army has never yet 'debated the reason,' but always obeyed its superiors; and they were right to do so." But in the end he returns to this, adding the words ad absurdum: "But it is laid down in the order of things: military despotism must inescapably lead to a revolution." But in addition we find, as indicated, much of extraordinary importance, such valuable scrutiny of the nature of the Russian people and vignettes of a psychological acuity perhaps never equaled in world literature.

Karl Nötzel: *Dostoievsky and Ourselves* (Musarion Press, Munich, paper 6 Marks, linen 9 Marks). Here again we take note of a fundamental work by Nötzel, in brief, surely the most important that has so far been published about Dostoievsky—and that this is of great importance today needs hardly to be mentioned. Nötzel assigns his book the subtitle "An interpretative essay on the man without preconditions," and he sees Dostoievsky as the model of a man without preconditions: "Seek and you will find yourself in every fellow-human whom you encounter. Learn to live as that person. And then—this above all—no longer conceal from yourself your inmost participation in the fate of your own kind. ... And that is surely the ultimate doctrine of this great master; acknowledge that you are obligated with the utmost devotion, with the most intense alertness, subject to no other than your own conscience, to work directly on the transformation of the established relationships among the whole of humanity. ... A man's veneration for his own kind should become the deed wherever a man meets another. ... Only by such an inward posture toward the environment and mankind can the individual consciously emerge from the inexorable chain of necessity, from the appalling compulsion of the war of all against all—and pass wholly to the domain of those who give of themselves. From all the bounty that has fallen to mankind, apart from the bare minimum needed to sustain his existence, he claims for himself only so much as he requires to preserve his ability to continue to serve his fellow-man. ... This we may call creative activity in its true sense. It can be accomplished only by the man without preconditions. ... You must no longer want to be anything other than a man in order wholly to be a man. But to be a man also means to be aware of infinite obligations beyond any accounting or requital. ... Dostoievsky was such a man without preconditions!"

Karl Nötzel: *The Destiny of the Poor* (Musarion Press, Munich, paper 6 Marks, linen 9 Marks). A brilliantly assembled "social breviary drawn from Russian authors," through which "the reader is brought spiritually closer to the fate of those of our fellow-creatures for whom we are all in some way responsible: the destiny of the poor, in all its incomprehensible depth." It includes passages from the works of Dostoievsky, Tolstoy, Chekhov, Leskov and Turgenev. Equally shattering is the picture painted by Leonid Andreiev in his narrative, published by the same press, *The Seven Who Were Hanged* (Musarion Press, Munich, paper 6 Marks, linen 10 Marks), about the last days of seven men sentenced to death for political offenses.

*Russian Critics* (Drei-Masken-Press, Munich, paper 0 Marks, linen 0 Marks [*sic*]). What are assuredly the three most important Russian critics are united in this selection: Belinsky, Dobrolyubov and Pisarev. We are here given a valuable and profound insight into Russian intellectual life in the century past. From Belinsky some choice observations about criticism and some actual critical notes, for example his famous "Letter to Gogol"; and from Pisarev some interesting obervations about Goethe and Heine. But I will conclude with Pisarev's words which are as significant for the Russia of today as they are for us, and once again underscore the importance of intellectual life (perhaps in a onesided way) for the realization of socialism: "But we have so far seen, and still see, two monstrous facts before us, from which all individual evils and adversities arise. First, we are poor, second, we are stupid. ... Of course, our spiritual poverty is no incurable disease. ... We have only to set it (the brain) in motion again and it will swiftly recover its natural strength. Of course we knew that, but the misfortune lies just in the fact that: we are poor because we are stupid, and we are stupid because we are poor. The serpent swallows its tail and is the symbol of eternity from which there is no exit. Charles Fourier is completely right in saying that the force of all the evils of modern civilization lies in this vicious circle. In order to be rich, one would have to improve, if only a little, the antediluvian methods of our agricultural, industrial and mercantile production, i.e., get smarter: but in order to get smarter we don't have time, because the poverty that oppresses us gives us no chance to breathe. But there is one possibility of breaking this charmed circle in two places. *First,* we know that a significant portion of the products of labor pass from the hands of the working population into those of the unproductive consumer. If the quantity of products that remain in the hands of the producer is increased, then his poverty is reduced and he is afforded the possibility of further advancement. This target was aimed at by the legislative measures of the government in the peasant question. It is at this point that a breach can be struck in the magic circle only with the help of legislative authority, and we will not dwell on this any longer. *Second,* one can influence the unproductive consumer, though naturally not with moralistic chatter, but by living ideas. ... This economy of intellectual forces is necessary everywhere and always, because mankind has nowhere and at no time been so rich in active forces as to be able to permit itself in its expenditure even the least wastefulness. So far that waste has everywhere and always been profligate, and that is why the results have to this point been so pitiful. ... But, in order to establish such an economy we must above all first attain the utmost clarity about what is beneficial to society and what is not. And for the acquisition of this clarity, it is precisely literature that must work. ... Once this objective has been undertaken, the magic circle will be breached."

A handsome edition of *Soloviev: Collected Works* has been issued by Der kommende Tag Press, Stuttgart; it will contribute to the dissemination of the writings of this profound mystic.

*Walter Landauer.*

[91-92 (Tenant Cooperative] omitted; for 93, see 88]

## 94. THE SYSTEM OF LIES

*Die Sonntags-Zeitung* (Heilbronn) IV (8) 2/25/23

In his article A *Socialist Reparations Plan*, published in the last issue of the *Sonntags-Zeitung*, Heinrich Ströbel identifies one of the most important and at the same time perplexing features of our "Unity Front"—namely the unity of deficient ideas and activity in the issue of an economic program. How very different, how much stronger would be the position of the Social Democrats if they at least could say: we have elaborated a certain economic program which takes account of all legitimate French demands and of German reparations obligations, so far as German economic conditions can permit them at all. Instead of that, things seem to be just as they were during the war: that ineptitude and inertia are hiding in the sanctuary of the hearth, and are difficult to impute to any one individual. But the press, including the majority of socialist papers, covers this situation with slogans and sensational reports.

It is an unheard-of sin of omission by the entire German daily press that the new French reparations proposals have been systematically withheld from the German reading public. This *Draft of a letter of the Reparations Commission to the Reichs-Chancellor* has been reproduced and discussed by the entire French press; it has likewise been communicated by the press in England and America. The German public has the greatest interest in gaining information about events of such importance abroad; and they would have every reason to react with the liveliest mistrust of a press which simply and wilfully treats facts of such significance with silence. This is once again exactly the same system of lies that our newspapers deliberately perpetrated during the war. And just as this system at that time did so much damage to us, to our cause and to our credit, so it will once again inflict the greatest harm on us. The objective of this system, now as then, is to place our intentions and aims in the best light, and to that end every variety of lies, exaggerations, distortions and suppression is recklessly employed.

This publication does indeed seem to me to be momentous. A critique of its individual points cannot be undertaken here, but it is in principle of extraordinary significance while an unresolved difference is pending between two countries, one of which has declared in great detail under what circumstances it is prepared to regard the dispute as settled. In my opinion, the publication of these conditions cannot otherwise be understood. For that publication would have been sidestepped altogether if France had regarded these proposals as superseded by events because of German resistance in the Ruhr. These conditions were and still are considered to be the basis of future negotiations. And it is perhaps also symptomatic that these conditions come significantly closer to our capacity to fulfil them than the earlier ones; perhaps it is precisely the intention of France to demonstrate to us its desire, even in the present situation, to resolve differences that are very unproductive for both partners.

Yet to conclude from this that France *will yield of its own volition* would be the greatest mistake with which one could delude oneself! I regard the policy of passive resistance, even from the viewpoint of "power politics" to be senseless, and am convinced that only positive actions—willingness to negotiate and offer counterproposals—will serve and save us.

*Walter Landauer.*

[95 (Hell); 96 (review) 97 (Guild Socialism; 98-99 (Russian cooperatives); 100 (Mauthner obituary) omitted]

## 101. WORLD DOMINION, RELIGION, AND THE KINGDOM OF GOD

*Die Sonntags-Zeitung* (Heilbronn) IV (32) 8/12/23

The end of past epochs and the beginning of new ones in the history of man are not so apparent to the observer as our historians like to represent on the basis of the dates of the reign of governing princes. Collapse and a new beginning are much interwoven. Hardly perceptible by its contemporaries, the new rises up. Perhaps this is happening today, but we are unable to recognize the delicate seed that will some day blossom into our hopes. It is in any case our duty joyfully to greet the miraculously rare, truly revolutionary actions and to serve their unfolding.

The rarest of such creative deeds are surely in the realm of "religious" life. Perhaps it occurs only at intervals of centuries that words of such austere criticism and at the same time of such promising constructiveness on this theme are heard as Leonhard Ragaz has offered in his book *World Dominion, Religion and the Kingdom of God,* Rotapfel Press, Munich. Not just the gaps left by science are filled up here, but the shallowness of centuries—above all those of our own time—is displayed, its dogmatism crushed, what was good in the old is taken as the foundation of the new. That said, we reach Ragaz' point of departure. His conception of historical development is characterized in the following words: "History is a continuous resurrection and not, as superficial scrutiny would believe, an all-engulfing grave. — In the name of the future we enthusiastically look to the past. For we discover that in the past, especially at the origins, truth is to be found which has later been lost to sight, and which we, however, must recover anew. We recognize in particular that the oblivious childhood of the race of men possessed a vision of the truth no longer recognized by the more mature intellect. — Primeval conditions of society become our teachers of new ideals. We eagerly ingest this old and oldest truth so as to create the revolution that is needed *now.* You surely see, dear friend, how superficial our accustomed contemplation of history is. It takes account only of a straight line, just as it erroneously deals only with rise and not with fall. We have to apply, if we wish truly to understand the flow of history, a formula that viewed geometrically would be nonsense; the line of development simultaneously runs forwards and backwards. — We are merely banal when we consider only the *one* direction."

And now I will attempt by a selection of passages from the book to give some indication of the power of his train of thought.

"The dominion is not a religion; it signifies rather the abolition of all religion. — Everybody understands by religion a quite specific spiritual structure that is not difficult to describe. What belongs to it is a doctrine—a doctrine about

God and divine things, a sacred doctrine. What belongs to it are certain rituals and morals, practices and institutions. Above all, what belongs to it is what is called divine service or worship. Also, some kind of community that forms about this doctrine and this service. What also belongs to it, moreover, is some kind of organization to establish this divine service, to regulate the community, to bequeath the truth that is confessed to the coming generation and to those outside. Some sort of formulated confession of faith cannot be omitted, even if it assumes the most unfettered form. Some form of partition from other religions is taken for granted. What is above all characteristic is that religion in this manner becomes *a self-contained structure.* It becomes a *structure unto itself.* It becomes an independent dominion. It separates itself from morality, art, science, politics. — And now we assert that what *Jesus* wanted is something quite different, more or less the opposite. — Not only did he not herald any *new* religion, but no religion at all. — It was not a religion that he wanted, but a dominion, a new creation, a new world. He wanted God, man, his brother, the new righteousness, the liberation of the world from fear and sensuality, Mammon, desperation, death and—religion! —What Jesus wants is a state of the world born of God. Nowhere does Jesus leave an opening to pious rites and works of any kind. — Nor did he establish any form of worship. —This is not mysticism, nor is it recondite talk of God. To adopt a phrase of Gottfried Keller's, one could say that his God radiates worldliness. Just as he reveals Himself in everyday things, so he is to be served in everyday things. — Jesus does not in the least aim at making people pious. He does not want religious people; his sphere is not that of religion. What he wants is not some mystical power separated from the rest of life. —The world is in truth his temple, thus the dedicated temple falls away. As a human being everybody becomes a priest, thus priesthood falls away; each day is holy, therefore there are no special holy days; every action is a divine service, there is no further need of some special form of worship; everything is sacred, therefore nothing remains that is specially holy or unholy. Religion falls away before God. — One can have much religion and still be far from God, even godless, really." "He (God) is therefore, wherever He appears in man, always synonymous with simple human morality, with goodness, humility, freedom, purity of heart, trust, above all with self-sacrifice, and vice versa: where these are, so is He, even if His name is not mentioned; where they are not, nor is He, even if His name resounds from every wall." — "Religion has the tendency to become something important for its own sake. — What matters is not to have religion, but to possess an essence, namely God's and man's essence. — Religion is the most powerful weapon of *reaction* — the dominion of God the most powerful weapon of *revolution.* — Religion is the most awesome force of history. — God's dominion parallels and surpasses religion throughout history." — "The Pharisees and scribes against whom

Jesus cried out are approximately what we today call theologians. They have established themselves on the throne of Jesus. Christianity, even Christ himself, has been theologized." — "There is something unspeakably sad in this posture of a great mass of people toward the church which they despise or affect to despise, about which they speak only with a derisory smile, and which they nevertheless cannot do without, especially when death approaches. — We encounter here one of the powerful roots on which all churches feed: the spiritual smugness, the lack of spiritual strength of one's own, and above all of real gravity in matters of the spirit. A representative, an abdication of effort to an institution, for cash or tribute, is much easier!"

"The church is organized religion, or religion as an organization. —The formula *extra ecclesiam nulla salus* [no salvation outside the Church] is not only the motto of the Roman Church, but uttered or tacit, the opinion of them all. A certain claim to infallibility is made by every church. But they thus become one of the most important sources of lack of freedom in the world. — They point *backwards,* they *repose* in possession of the truth, they are *beata possidens* [possessed of blessedness] and thereby of course constitute a powerful inhibition to truth and freedom. — And thereby access by the masses to the Almighty is made impossible. The masses are denied access to the most important and decisive tasks which are transferred to a small circle. Conscience is replaced by authority." — "That is also the answer to every objection advanced by the defenders of the church; what would happen if it were not there. To them we pose the counter-question: "Is an *illusion* not worse than a *void ?* —Would it not represent progress if the void were not at least made clear and painful?"

"Our commitment to social democracy was a commitment to the cause of socialism, to the guilt of society, especially of Christianity. It was a protest and a confession of faith. Our path to social democracy was not the path to a political party, but a path to the followers of Christ. Religion is in league with the ruling powers, the dominion of God is surrendered to publicans and sinners." — But "social democracy, from being a movement for the dominion of God, has become a religion. — The party lost its sense of the ultimate goals for which it really stood, and became a cause concerned only for itself. — Just like a church it (socialism) did not heed whether it was a community of those with conviction and determination, but rather that it should gain the largest possible masses — social democracy became a church. All the signs of one have appeared: the organization that has become important in itself, the claim to infallibility which denied to the individual his own search for the truth, something it would rather not see. The priests come and the dogma and the holy scriptures and the confessions of faith, and naturally their exegetes, the theologians and scribes, and then the heretics and the inquisitors."

Ragaz rejects dogmatic Marxism; likewise the class war in the narrower sense in which party propaganda uses it. "Where force governs, socialism does not, i.e.: free fraternal community. For violence is domination. Socialism that seeks to come to power through force founders in advance from an internal contradiction and must succumb to the truth of this adage: 'Every empire and every house that is disunited must collapse.' — Our socialism has room for this meaningful, creative and superior action of the individual, of the prophets, apostles, heroes. It is dependent on them. It is not only reconcilable with a correctly understood individualism, but is even inextricably linked to it. In this relationship what rules is not conflict but polarity. True community life gains among the masses as true individual life increases, and vice versa. — For what is a socialism that, as socialism, did not trust the individual as individual? Any socialism whose boundaries are more limited than God and man, is too little for us. We do not have the least desire to sell our free souls to it. — Our socialism is in sum and substance a socialism of freewill. – Just as in our 'religious' life we want no 'church,' so in our social life we want no 'State'; we want a *free community*. — The community in our sense must be from its inception *socialistic* in intention. It may not be an imitation of capitalism, may not be based merely on some ingenious calculation of utility, but must proceed resolutely to be a community of people who wish to live with one another in a new, truly human fashion, first perhaps in economic relations, but in the conviction that this is only a point of departure for a new ordering of our *whole* lives together. — The consumer cooperative must be linked to the *production cooperative.* —

"The glorious task of the working cooperative is in any case to bring about a *rebirth of labor.* — Each and every village must become a free socialist cooperative, as the ancient villages often were, even if in incomplete fashion. These cooperatives in the country and in agriculture must enter into a relationship with those of the city and of trades that is not one of hostility, but of barter, complementarity and assistance, and so close the gap between city and country, peasant and worker, as to construct a real people's community. — *Federalism* (which is after all only another word for free cooperation) will be the shape of the future also in the organization of peoples. — It may be a violent social *revolution* which, growing out of the misery and injustice of today's world, shatters the existing social order, a harbinger of a new spirit. When it comes we shall not condemn or fight it, perhaps even greet it, not assuredly as the best, but as second-best since we shall perhaps under certain circumstances collaborate in compulsory legislation. But we shall always know that this is not *our* socialism. Our *actual* work is the annunciation not of the 'law' but of the 'evangel,' the advocacy of spirit and of freedom, for the construction of a society, of a world in the free love of God and man. — Just as Christ is the end of church and religion, so is He also the end of the State and of politics. —

Wherever freedom and love reign, State and the right are annulled. — Hence no church organizations and complicated rites, but free and diverse groups of people who aspire in free relationships to the one great end—that is our ideal of a *real* religious community'."

This is a fragmentary selection of the declarations of particular interest to us *here*. With difficulty I decided to take the separate passages out of their context, but in the end it seemed to me still better than being obliged to replace them with less perfect passages. I had the further hope that this method would increase the desire to read the book itself. Whole chapters—and not the least important—could not be mentioned at all, for example, the demonstration that the loss of "love" is the most profound cause of the condition of our world, and that salvation through love is the only path to rescue. Of the greatest importance for an understanding of our time are also the explorations of the Reformation, Catholicism, paganism and Judaism, from all of which the little good they have in them is separated from the weeds that flourish around, and weld them into a powerful synthesis. But I cannot say anything other than: read and come to terms yourself with this shattering accusation against our present existence, this radiant call to a new community.

*Walter Landauer.*

[102 (reviews), 103 (Model settlement) omitted.]

## 104. JOSEF POPPER-LYNKEUS AND THE *GUARANTEED SUBSISTENCE FOR ALL*

*Der Aufbau* (Zurich) IV (33) August 17 1923

Our century has not looked too kindly on social and socialist *programs* but is predominantly busy, when it addresses Socialism or social questions, with national economic doctrines and with the struggle about economic "laws," or—in the socialist parties with current events of the day in politics and parliamentarianism. How much the workers' parties have come to grief on the latter aspect, to what degree they have lost sight of their great objective and have prostituted themselves to inactivity is shown by their helplessness during the war and, more recently, in facing the problems of reparations. I call their actions passive and senseless because, however sincerely intended, it is only a rearguard action which leaves them at any given moment baffled, without a program—without a Utopia—in the face of reality. And those who call themselves socialists have not only failed to contribute bricks for the structure of socialist ideas, but have rather, by their systematic silence, prevented the few great and constructive efforts from becoming effective.

Among these creative achievements I count the program for a *Guaranteed Subsistence For All* (C. Reissner, Dresden) by Josef Popper-Lynkeus; the purpose of the following lines is to reiterate his principles here.

Popper aims to solve the "social problem" with his proposals. Thus we must first grasp what he means by this question. He says: "By '*social problem*,' I mean the problem of finding an institution suited to assure each individual of the *necessary* economic means to sustain life, without making this dependent on the will of other people." Popper thus proceeds in his definition of the social problem from ethical presuppositions and assertions which he has addressed in greater detail in a specific work.[1] Its postulate is the greatest regard for the unconditional acceptance of the existence of each and every individual, and it is on this that the doctrine of the *guaranteed subsistence for all* is founded; just as it is for example expressed in the words of the *Talmud*: "For the sake of each individual human being the world was created," or "The soul of each man is worth more than the whole world." With this respect for each existence (for the rights of man that lie in his having been born, in his being alive, as Nötzel says), for the fact that "an individual is something that does not want to cease to be, to continue to be what it is," as Popper says. With this is connected the recognition of the equal value of all individuals as concerns their need to be assured of their economic necessities. The highest maxim is "The greatest possible economically necessary security of life for *everyone*" (in contrast to Bentham's principle of the greatest happiness of the greatest number). The State—if we use this word, it is in the first place only as a synonym for the totality of citizens, without any statement about the nature of their interrelations, roughly as Wittel's State is made the equivalent of the words: "we as society" (*The Elimination of Need*, Anzengruber Press, Vienna)—takes on the obligation to take care that within its borders every individual, without exception, receives all that is necessary for the maintenance of life, that the minimum for existence is guaranteed. For this, according to Popper, it is not only necessary that "everything needed to meet the necessities of life be produced or in some fashion procured by the community, but that it is equally necessary that these products remain within the authority of the community, thus to be held under its control so that it can be distributed in equal measure to all its citizens." The acquisition, i.e., the enjoyment of all the goods for which the individual, with

[1] *The Individual and the Evaluation of Human Lives.* Further materials are to be found in Popper's works: *The Right To Live And The Duty to Die*, *Fundamentals Of A New State Governance*, and *Voltaire*, all C. Reissner, Dresden. *Autobiography*, Unesma Press, Leipzig.)

the best of wills, cannot wait, may not be conditional on anything other than *on their availability*; for "it is nothing but a cruel game to measure necessary goods, not directly, but only in accordance with some—always merely conventional—scale of values, in accordance with some economic 'thing-in-itself,' for example, time worked or sums of money, and then only in exchange." All protective measures applicable only to certain classes or segments of a nation are worthless—only the protection of all within a collective economy makes sense. The repeatedly calculated and discussed figures on gross national product put forward by national economies organized like corporations, and the relative standing of this wealth in various countries, are worthless playing with numbers as far as our problem is concerned. A statistical examination only makes sense to determine whether a particular state is in a position to secure the necessary decent standard of living for all its members. Scrutiny of the causes of the incurable sickness of our economic system shows that the social problem (in the sense defined above) can be cured only by the distribution of the minimum for existence in actual goods. Popper points with special emphasis to the significance of the private crises which demonstrate that it is "no less ridiculous to try to restructure our economic system (in theory) *by means of* unilaterally paying attention solely to the situation of wage-earners, and pedantically deploying what in any case is a disputable subtlety, as Marx has undertaken with his concept of 'surplus value.' It is the fact of surplus value itself that needs to be demonstrated or explained in light of the misery of our economy where we encounter, day in day out, innumerable incidences of poverty and need in so many classes of society, not just the wage-earners, where the question of surplus value does not arise at all!" The origins of these facts lie overwhelmingly not in public economic processes, but in the chronic and acute private crises.

We turn now to the principal question, namely how, according to Popper, the minimum for existence is to be secured, what it should include and how the rest of the economy functions. In order to assure every citizen from birth to death with complete freedom from need and distress (as concerns his own possibility of existence) we must include in the minimum for existence not only food, housing, and furnishings, clothing, heat and light, but also medical assistance, nursing, medications etc. These necessities are to be produced or otherwise acquired by a fraction of the society working as an obligation, and their production and output must be distributed or guaranteed to all citizens, without exception and unconditionally, not in the form of money, but in real goods and services. This segment of the populace with an obligation to produce is called by Popper the "Labor Corps." It works for itself and for everybody else. In addition to the specified primary or *survival* necessities, there are also secondary or cultural needs (books, theater, travel, etc.). These secondary, but also necessary needs should, as far as possible, be likewise incorporated into the standard of living and be supplied. A plebiscite will have to be conducted from time to time to set the limits of what constitutes luxury measured against the statistically determined performance of the minimum organization.

The complex of arrangements serving to provide the primary and secondary needs is called the minimum organization. The primary part of that minimum is produced by the labor corps as real goods; the secondary part is merely guaranteed, *not* procured by a public *agency*, but distributed in the form of money, and purchased for money within the continuing free economy. The secondary needs cannot of course be met by direct allocation, because they are intrinsically diverse. The procurement and distribution of the minimum is overseen by an independent authority, isolated from all political influence—the "Ministry for Living Standards," to be elected by popular plebiscite. How Popper conceives of this election is not clear from his account.

The decision that in the area of the necessities of life everything is to be distributed in goods, and must exclude all use of cash, is unavoidable if one is really to avoid all the uncertainty of life today; only thus can even the attempt to speculate with items of daily need be precluded. The money needed for the distribution of the secondary minimum is to be raised out of the proceeds of state monopolies, levies, tax monies, etc.; its level should as far as possible correspond to that part of the income of a comfortable family that is not applied to subsistence needs, but it must of course be set within the limits of practical possibilities.

Following the introduction of the minimum organization, the economic structure of the State would be as follows:

1. The most important part of the public economy will be represented by the *Labor Corps* as a duty-bound (or compulsory) economy for the procurement and distribution of the primary minimum; this involves no money dealings, everything is distributed in goods. Purchase, barter, or speculation do not take place here; the monstrous cost of the unproductive activity of middlemen is saved; likewise insurance costs will for the most part be eliminated.

2. The *free State economy* includes state monopolies like post, telegraph, railroads etc. Service in them is voluntary; wages are paid in money, but because the minimum subsistence is assured (primary needs) they are naturally lower than at the present time.

3. In addition, there would also continue to be a *free private economy* with money as the medium of exchange and with free competition; this can, however, involve only production and consumption of goods and services that are *not essential;* this is fundamentally a *luxury economy.* This would in large measure avoid at least the currently disastrous consequences of a capitalist economy.

For the need to preserve a free private economy and for the importance of the separation of the necessary and the unnecessary, Popper lists a long series of reasons; the most important are:

1. It would be much too complicated for the production and completely impossible for the distribution of luxury goods or services to implement a collective system based on the equal rights of all.

2. Without this separation many more people would have to be enrolled in the labor army; on the other hand, compulsion should be exercised only within the limits unavoidable for the security of each individual.

3. In the absence of a free economy the stimulus toward technical progress and invention of every kind would be lacking.

In this manner the connection is maintained among the security of the economic existence of all and the *equality principle* applicable to each individual existence as such, *freedom* in economic respects so long as the economic security of others is not thereby threatened, and progress in the public economy, and technical and other advances are not disturbed.

In order to assure the minimum organization Popper regards the following measures to be particularly necessary:

1. Within the minimum organization there are to be, as already mentioned, no money transactions of any kind; otherwise new uncertainties are unavoidable.

2. The minimum organization must be as independent as possible from colonies and foreign parts. This provision will be superfluous if there are international organizations in the future.

3. Everything that is distributed by the minimum organization or is connected with it must stand under the protection of international law.

4. A certain stock of all minimum goods must always be on hand.

5. Further ordinances (a general reduction) will be needed if for any reason an insufficient supply of provisions is produced.

6. All other authorities in the State must be subordinate to the minimum organization or its highest officials; it must always have the final decisive word because it is responsible for the assured existence of everybody.

Anyone will be free to decline his share of the minimum if, for example, because of receipts from free private enterprise he disposes of a sufficient income. But under no circumstances does this entail exemption from compulsory service in the Labor Corps. (To be continued)

*Walter Landauer.*

## 105. JOSEF POPPER-LYNKEUS AND THE *GUARANTEED SUBSISTENCE FOR ALL*
(Continuation)

*Der Aufbau* (Zurich) IV (34) 8/24/23

We turn now to the main principles of the constitution of the Labor Corps. All persons of specified age-groups, male as well as female, must serve; only sickness or physical unfitness may disqualify from service; there are no preferential reductions in the time to be served. Physical and mental qualities, such as aptitude, should be taken into account as much as possible in making assignments (i.e., free choice of activity within the limits of quantitative minimum needs). In taking account of professional, trade or other education, entry into service may be delayed, or half-day service can be given (naturally then for double the period). After their period of service is over each person who prefers further service in the labor corps to unemployment or activity in the private sector is free to work in it, but now for appropriate compensation. The number of daily working hours will be set depending on the nature of the work. Housing, clothing, etc., will be supplied, while avoiding any kind of uniforms. When traveling the subsistence minimum will be supplied everywhere on displaying an identity card. The calculation of the labor corp's activity will be based on the total population, less those people who have declared a specific renunciation for the coming year. Thus, obtaining the right to draw supplies is not dependent on any registration or other formalities; in this way the *entitlement* to the subsistence minimum is emphasized, and any appearance of welfare or charity avoided.

In a section nearly 200 pages long, Popper undertakes a statistical examination of the functioning of the Labor Corps and of the ways this will work out. We cannot go into the thoroughly interesting details here; only the conclusions are repeated here; these all apply to Germany: "To be able to supply all citizens with food, housing, clothing, as well as medical and nursing care with complete certainty throughout their lives, it would be necessary to have the consistent labor of a subsistence or minimum army; about 7.75 million men would be obliged to serve beginning with their 18th birthday until their 31st, i.e. 13 years, and about 5 million women from 18 to 26, i.e. 8 years. After completing this period of service they would be completely free and can spend their time any way they please. Their daily working time would never exceed 7 to 7½ hours and would be further reduced depending on the difficulty or danger of the work involved." According to Popper's calculations productivity within the minimum organization would increase over the present—but even if it were lower that would not be a reason for rejecting it, because "if there is to be starvation, then let everybody starve equally."

Let us note finally that the concept "minimum" should by no means be understood to imply so restricted an access to life's necessities that it could not be less; rather the minimum is to be *above* bare necessities, and in shaping the society of the future, the constant goal should be the *maximum possible* and not an actual minimum.

I have made the attempt to repeat the principles of Popper's subsistence minimum program and hope that I have to some degree made its most important aspects clear. I have already noted that in doing so, many very significant details, even whole sections have been left untouched, for example, the *Statistics concerning the Labor Corps* and the *Transition from the Contemporary to the Socialist Economy*, and the excellent critical argument about the irrationality of all previous economic doctrine and social politics. To point once again to the fundamental motif of Popper's teachings, I need to repeat only the title of his earliest socialist work (now unfortunately out of print): *The* Right *to Live and the* Duty *to Die*. In a word: unconditional proclamation of the assurance of every human life; the guarantee of a minimum subsistence, and elimination of compulsory military service.[1] It is these principles that assure Popper's life work of its significance and relevance until the definitive actualization of his proposals.

A few critical observations and some comments now follow on my thoughts about how the *Guaranteed Subsistence For All* would fit into cooperative socialism.

All substantial objections that I have to lay against Popper's exposition in the last analysis are grouped around one problem best characterized by the words centralization or decentralization, State measures (or measures of "popular representation," if you please) or self-help. Popper's views are characterized by a certain conscious bias *vis-à-vis* socialism and what he calls social problems. Popper is surely correct when he sets his face against "total socialization"—to the extent that it is indeed unfeasible at the present time—and instead pushes to the foreground something that is at least *more practicable,* if only all the interested parties could ever agree on a single agenda, namely the assurance of sustenance of life. But it is beyond question that any change in principle of our economic lives is of significant influence on the future formation of community life, or at least could be. It is therefore inappropriate to alter those economic conditions without taking into account the possible consequences of such change; the weightier and more incisive those changes, the greater the responsibility toward future generations. Even if the possibility

[1] (Popper's insistence on "voluntary military service," which is argued above all in his book *War, Conscription and State Constitution* (Rikola Press, Vienna), cannot be pursued here; I will attempt to expound critically in another place [see next article] why Popper's suggestions constitute only a very unsatisfactory solution of the problems in question.).

existed of implementing Popper's plans overnight, we should still face the decision whether, in order to assure people's existences completely, we should accept a social organization that does not point in the direction of our future plans and hopes. And probably we would decide *in favor* of Popper. But that is nonetheless by no means the case. Popper himself reckons that his ideas will not attain the stage of serious consideration before the year 1980. Even when that moment in time comes closer to us it would still be imperative to take under consideration whether it is not possible, at the same time as realizing all the positive aspects of Popper's program, to attain assurance also of the more remote aims of social development. In fact we believe not only that that is possible, but even that all the provisions protective of the single individual could be placed on sounder ground. (To be continued)

*Walter Landauer.*

## 106. JOSEF POPPER-LYNKEUS AND THE *GUARANTEED SUBSISTENCE FOR ALL*
(Conclusion)

*Der Aufbau* (Zurich) IV (35) August 31 1923

Popper says at one point: "Who then is the State? We are ourselves the State ..." We must dwell for a moment on this fundamental mistake. Constantin Frantz has *proven* in the most dramatic fashion that all forms of contemporary parliamentarianism are incapable of executing the will of the people, even that in *every one* of their decisions a minority is victorious over what in some circumstances is a substantial majority. He has verified the demonstration by reference to all historical experience that it is embedded in the essence of the centralized state that the interests of the single individual are ignored or even violated. It is, therefore, clear that today as in every centralized state the "State" is not "us," i.e., the separate individuals, but that it represents an administrative apparatus strongly influenced by the ruling class, effecting a compromise between the efforts of its officials and the government emerging out of more or less fortuitous parliamentary coalitions. But we want to construct a state in which we and the interests of each individual are reflected, so far as the general welfare permits. We must, therefore, seek to assimilate the ideas of the *guaranteed subsistence for all* to a decentralized community structure. Popper says: "Every concentration is the more effective, the wider it extends. Community socialism may operate only as a subordinate part of state socialism." What does that mean: "the more effective"? It refers to the authority, the power incorporated in the collective which Popper calls upon in order to secure his demands and to execute them. But surely nothing is more risky than this "assurance"! Constitutions that entrust the protection and specification [of rights] to a central power have nonetheless been constantly disregarded—and the people have always allowed themselves to be misled into surrendering their highest rights for much less and much worse than a mess of pottage. We assert that the true security of the single individual is assured only by the converse alignment of authority, and we simultaneously assert that it is just that which embodies the deepest sense of Popper's ideas. Thus we establish, as the first result of our inquiry, that we embrace Popper's leading thoughts as perhaps the most important constituent of a constitution, the binding principle of which is a federation of unions of self-governing communities.

Popper proclaims a universal obligation to work, but at the same time limits it to the degree required to assure the absolute necessities for life. We agree with that, but we demand that in performing this obligatory labor the life of the individual not be subject to a central authority over which he has no influence and which is ignorant of his interests and life's needs, and which limits him and threatens his capacity for development. In his *Fundamentals of a New State Governance* (C. Reissner, Dresden), Popper prefaces the requisite improvement in legislation with the motto: "For secondary needs, the majority principle, for basics, the principle of guaranteed individuality." We reject the current usual majority principle (parliamentarianism) and aim to secure the rights of the individual to a higher degree by his membership in small communities.

Popper explores the ideas of cooperative socialism only very briefly. In the great majority of contemporary consumer cooperatives it is indeed the case that they "actually embody nothing socialist," because they aim "only at greater assurance of goods of superior quality and lower prices by independence from middlemen." Popper sees only what the cooperatives have become and not what they could become. A few lines later he continues: "Let us imagine such a consumer union (with production facilities) extended to *all* citizens, with all its articles produced by union members, as to some extent is already the case (in England and Scotland recently) and instead of selling their needs to members (and their families) *distributes* them directly in the form of goods, *then what we have is my program precisely realized.*" Much in that is correct. Above all, it is a fundamental principle of a socialist cooperative that the members must themselves be the producers, and not that the production, as is the case in most of the plants in the contemporary cooperative system, is manufactured by employees of the cooperative. Furthermore it is a basic requirement for the cooperatives that they dedicate their production in the first place to articles necessary to life (the minimum needs), but they should be free to extend their activity beyond this to the areas that Popper wishes to reserve to the private sector. Once again we must raise our earlier objection: why is it necessary for all citizens to be united in a *single* cooperative (State cooperative), and only in this? We also propose a State cooperative as the final stage in the unification of all cooperatives—but as concerns the labor corps we would assign to it only those tasks that would correspond approximately to the duties of an oversight committee of the union of self-governing cooperatives: thus, for example, supervision to assure compliance with the state cooperative constitution, as a court of arbitration for matters contested between cooperative unions, State cooperative statistics, etc. In contrast to this, the productive tasks of the labor corps and all duties arising out of these would be assigned to the separate cooperatives. In the military a concentration of the population in garrisons runs parallel with the centralized administration; there is no question of that in the case of the labor corps; our insistence on decentralization therefore applies solely to the administration; the problem of subsistence mandate and decentralization is not posed as "either/or, but as "one as well as the other." It is clear that in speaking of our organization we can no longer refer, as Fritz Wittels does (*Die Sonntags-Zeitung,* Heilbronn, No.10 of March 11 1923), to an "army under labor discipline"; that is just what we want to avoid. And it is Wittels himself who, in a passage in his quoted book, in a brief allusion, amends or develops Popper's ideas, in our direction. He says: "It is possible that the "leaderless" [anarchists?] are right when they say that the State (of today) is incapable of leading everyone to security. Then the cooperatives will think about developing alongside the state, will unite and eventually constitute a flourishing state within the expiring State. They will become the state, or the economic community. What then remains of the State might be of interest to the politicians, the professors of law, the cultural historians; but is of no interest to us on the road to solution of the stomach question. The subsistence army could also be formed on the way to the cooperative—a detour." We can only add: a *necessary* detour! Popper himself supports in another book (*Fundamentals of a New Political Constitution*) [see bibliography] the vigorous promotion of community socialism, but even there he rejects autonomous socialist communities because he believes that "the organization of the labor corps *throughout the state* and its geographical distribution can be accomodated only in relatively small measure, or at least not completely, by the existing communities." We believe that this accomodation by no means poses insuperable difficulties and that it could be managed. From a different direction Popper quite clearly sees the limits that are imposed on the possibility of centralization altogether. Thus, for example, there can be no question of incorporating within his labor corps either small industry or the physicians, and he adds to this the acknowledgement that a "centralized administration is not at all the objective of a true social reform, but only a useful means in those cases where it happens to fit." We say that this centralization is in no case useful, and that the whole minimum organization could be assimilated just as well in a decentralized community system. In the pamphlet *Defense Of The Guaranteed Subsistence Program* (Press of the Guaranteed Subsistence Propaganda Office, Vienna) the following observation is to be found: "It is a mistake to believe that concepts like centralization, nationalization, socialization, concentration, cooperative, organization, etc., constitute the essence of socialism. These methods are always *only a means, never ends in themselves;* they can be used when they are useful for the assurance of a standard of living; but one must not allow these terms to become a prescription nor allow them to turn into a cult." That is, of course, correct, and it is only in the choice of these means that we differentiate ourselves from Popper. Questions of criminal justice and hereditary rights, of education, etc., *may* indeed as such be separated from the introduction of the existence minimum, as Popper does, but even if one joins with Popper in the discussion of this program in abstaining "from speaking of the social life of the entire [society]," it remains the case, on introducing this new economic form, that at the same time the foundations for it be laid, so that in this higher economic stage a high form of law, morals, science, and culture can also flourish.

Popper surely underestimates in a high degree the significance of "self-help" (Schulze-Delitzsch) as well as the dangers of bureaucracy. All centralism generates an elaborate administrative apparatus, and the danger of bureaucratization surely increases with the numbers of officials needed. Self-help, however, is one of the most important bases of any new organization insofar as insight is gained into the utility and need of new collaborative structures; and communities function in a much more secure and practical fashion than do compulsory organizations.

In the case of agriculture Popper indeed recognizes the value that cooperative organization has already produced here and there, but even here he still prefers the large collective undertaking. We believe that both forms of agricultural operation eliminate the disadvantages of small farming (cf. David: *Socialism and Agriculture*, 2nd. ed., 1922) and that cooperative organization makes all the advantages possible that Popper wants to supply in his program. The "multicolored variegation of cooperative structures," which Popper criticizes, seems to us one of the benefits compared with Popper's program because it leaves more room for play in further developmental possibilities.

I do not want to go into objections that can be raised against many details of Popper's program; but will refer only to two more of these: first, that the opinions of the *International State Right of Inheritance Propaganda* have many connections to Popper's program and each complements the other in many respects; I will shortly return to this in a pamphlet. And then, that the retention of personal property in no way presupposes a private sector economy, as could appear to be the case in Popper's justification of the necessity of a free private economy.

"The moral progress of mankind has so far principally emerged from its institutions, and not from the private morality of single individuals." This observation is Popper's point of departure. He too wishes to change the institutions, without presupposing other people, new insights or some higher level of knowledge. All that is required for the realization of his program is the consent of the people. He sees the solution of the social problem in securing existence and in the freedom in large measure to shape one's happiness in accordance with one's own wishes. This is what he believes himself able to prophesy as the outcome of his revolutionary proposal.

*Walter Landauer.*

[107 (Mahatma Gandhi) omitted]

Professor WALTER LANDAUER, Heidelberg:

## 108. JOSEF POPPER-LYNKEUS AND THE *GUARANTEED SUBSISTENCE FOR ALL*

*Monistische Monatshefte* (Hamburg) IX, February 1924

[This is a redaction of the three articles (No. 104-106) in *Aufbau,* to which Landauer refers. It is mostly expository, with the bulk taken either as direct quotation or close paraphrase from #104 and #105. Only the concluding paragraphs are included from the otherwise partially critical review of Popper's views in #106.]

These are the principles of Popper's program. Many details, important in part, have had to be omitted here. The unlimited security of each human existence by the guarantee of a minimum—"the right to live"— and elimination of war service—"the duty to die"—is what Popper demands. My thoughts about Popper's program as incorporated within cooperative socialism, if in the place of a centralized structure of the minimum organization there is a decentralized one, among other critical observations, I have set out in another place (*Der Aufbau*, Zurich 1923, Nos. 33-35). Popper himself says in one passage: "The labor corps could be based on a cooperative: this would be a detour." And this, it seems to me, is a necessary detour.

"The moral progress of mankind has so far principally emerged from its institutions, and not from the private morality of single individuals." This observation is Popper's point of departure. He too wishes to change the institutions, without presupposing other people, new insights or some higher level of knowledge. All that is required for the realization of his program is the consent of the people. He sees the solution of the social problem in securing existence and in the freedom in large measure to shape one's happiness in accordance with one's own wishes. That is what he believes himself able to prophesy as the outcome of his revolutionary proposal.

[109 (Gender Psychology) omitted]

## 110. HENRY FORD

*Der Aufbau* (Zurich) V (20) 5/16/24

Henry Ford, the American automobile manufacturer, recently published an account of the rise of his enterprises and of the ideas by which he was guided (*My Life and Work*, Paul List Press, Leipzig), that tell much of interest and on critical scrutiny give rise to a number of thoughts. The book seems to have stimulated much interest and much false appraisal. Nothing is so characteristic in this as the unison of judgment in which enemy brothers find themselves in agreement: The *Deutsche Allgemeine Zeitung*, an industrial paper by the grace of Stinnes commends the book, saying: "When one puts down this book, unusually rewarding in its thinking, one asks: isn't a world of ideas opened up here that could also help our German people to recover?" and the *Vorwärts* (Berlin) adds the complement to this: "One cannot help wishing that German entrepreneurs, with Herr Stinnes at their head, would take instruction from Ford and familiarize themselves with his principles." Could the miracle have occurred that industry and *proletariat* have united in a common social program? Has Ford found the way to unite the interests of capitalist industry and the demands of socialism? In fact Ford's really tendentious narrative aims at displaying the organization of his industry as the point of convergence of the interests of employers and employees, in which there is an end to all class warfare, and each is a satisfied member of the productive community.

The fundamental thought on which Ford's undertakings are based can be summarized in a few sentences. Ford saw that big business could be attained with automobiles only if there is a large market of buyers. All the production methods that Ford employed over the years looked back to the question: How is it possible to obtain massive turnover with automobiles? That is possible only if the vehicles are cheap, much cheaper than all other makes, but still they must be of good quality—in a word, "the approach of the business must always have the performance of a service in the foreground of consciousness," and Ford rejected cutthroat competition. The goal of low price and high quality led him to reject the manufacture of the greatest variety of vehicle types, as was customary elsewhere in the industry—thus Ford built only one or two models. That made possible mass production on a large scale. Even so, lowering the price soon found its limits. What next? Ford mechanized the work in a hitherto unprecedented manner. He formulated the ground rules to be followed in the assembly of automobiles and of individual components in the following way:

1. Array both tools and workers in the sequence of the processes to be undertaken, so that each component of the assembly has to move the shortest distance.

2. Make use of chutes or other means of transportation so that the worker on completing assembly of the component on which he has been working always stays in the same place. Make all possible use of gravity to bring the item in question to the next worker.

3. Make use of the assembly line to deliver the component parts at convenient intervals to and from the worker.

The whole structure of Ford's enterprises rests on this foundation. As an example of the success of the mechanization of labor Ford mentions that the assembly of magnetos originally took *one* worker 20 minutes, while after the work had been broken down into 29 separate steps and with the increased use of an assembly line, only five minutes were needed; or, the stationary assembly of a chassis took 12 hours 8 minutes, but required only 1 hour 33 minutes after the work had been broken down and placed on an assembly line. The consequences of these savings of worktime are of course extraordinary. Let us consider the most important further conclusions that Ford drew from his production system. Originally the whole automobile was assembled in a *single* factory, but today the Ford works consist of plants making automobile *parts.* "The machines are more closely placed than in any other factory in the world—in accordance with a system that gives each worker every square inch of space that he needs, but as far as possible (so as to avoid higher production costs) not a square inch more." Finally, there comes the vertical organization: raw materials (mines, forests), railroads, forges, parts factories, assembly, sales, and repairs.

Now that we know that the cardinal principles of the Ford works are cheap products and a large turnover, we ask with some trepidation how is it with the wages. Well, this question is also answered by Ford consistent with his system: "No question is as important as that about wages—most of the population lives on wages. Their standard of living and income level determines the welfare of the country. Thus says Ford, and sees it as the goal of each worker that he can drive to work in his own automobile. Indeed, he goes further and says: "It is not the custom to characterize an employee as a partner, but he is nothing else ... It is complete nonsense that capital and labor view each other as separate parties—they are associates. If they work against one another, if they pull in opposite directions—all that does is to damage the organization in which they are still both participants and to which they owe their living." The fact that "in all Ford plants a minimum wage of $6 per day has been introduced" is still astonishing even if one takes into account that the purchasing power of money in America (measured by German standards) is only about 45% of face value. Just the same we had better note here that in Ford's book, besides this unconditional assertion of a minimum daily wage of $6, the following puzzling and contradictory passage is found: "Today we pay *without any restrictions whatever* (My italics. W.L.) a minimum wage of $6 per day ... Each worker receives his minimum wage *as soon as he has reached the minimum output* in his productivity—and that depends exclusively on his willingness to work. ... That people really work is evident from the fact that about 60% of the workers earn more than the minimum wage. Thus $6 per day is not the average but the minimum wage." Let anyone understand this who can; I cannot—and I do not think Mr. Ford does. And that makes one somewhat skeptical about Ford's statements altogether. All the same we need to turn to causes of greater concern. To the credit of Ford's plant management, there are low prices based on maximum mechanization of the labor process and assurance of high wages; is there no debit corresponding to this? If that were the case, then one would of course approve Ford's system in the interest of the customer as well as that of the employee. I would have attached no importance to what an American socialist told me recently, that the whole State of Michigan (in which the Detroit Ford plants are located) "curses Ford, because he has ruined the population"—if Ford himself had not supplied convincing evidence for the truth of this assertion. Bear in mind that the workers active in the Ford works are regarded as "associates" and receive record wages; their only desire would be to stay there and accumulate a small fortune. Let us hear from Ford himself: "In 1914, when the first plan (for a raise in wages) went into effect, we had 14,000 employees, and it was necessary to engage 53,000 men in order to maintain the work force at the level of 14,000 men. ...Now the turnover in personnel, as far as we know, runs about 3 to 6% monthly." And Ford cynically adds: "The problem of personnel turnover has not given us any headaches since then. ...We do not even keep any statistics any longer in this area because the personnel turnover question is of such little interest!" The figures that Ford himself gives, and about which, from the manner in which he shrugs them off, we must assume that they are higher rather than lower, are unprecedented in the history of major capitalist industry, and give us an idea of the hellish fate of workers in the Ford works. Anyone who knows under what appalling conditions the workers remain on the job, even though rather favorably paid, will need no commentary on these numbers. It is surely beyond all doubt that any worker, as soon as it is at all possible for him will leave the Ford works. And for this Ford himself gives an interesting example. In 1919, some 30,155 instances of personnel changes were recorded (among an approximate total of 25,000 staffers). "In 10,344 cases the individuals were absent for more than ten days without giving notice, and so were struck from the list of employees. For refusing assigned tasks or as a result of transfer requests without giving more precise reasons, 3702 were laid off ..., some 3000 were transferred to other factories. About the same number wanted to go home." For what reasons the other 10,000 men left, we are not told. From these figures it is abundantly clear that the productivity of the Ford works along with the manufacturing methods, and just for this reason, i.e., the manner of carrying it out, depends in the first place on the most ruthless exploitation of the productive capacity of the workers. The human machines who are posted on the Ford assembly lines are subjected to intolerable pressures—they succumb to it, or escape. All those who admire Ford's industrial organization (and among them are Foerster in *Menschheit*, Schairer in the *Sonntags-Zeitung*, and Flake in *Weltbühne*) have overlooked the sacrifices which yield these successes.

Ford believes that it is necessary in defense of his system to establish that abnormal consequences of the extremely restrictive activity on the physique of his workers (nothing is said about their spirit) "could not be diagnosed in a single case." Moreover: "*At their request* our people are *regularly* transferred; it would be a small matter to arrange this if that is just what they wanted." Thus they simply love their jobs! As an illustration of the reliability of Ford's data one should recall that he elsewhere said that in a *single* year "for refusing assigned tasks or as a result of transfer requests without giving more precise reasons, 3700 men were laid off"!

Now it will be said in rebuttal to me that Ford's book all the same contains some very sensible ideas. True, and I will list some of them: "Business is nothing but labor. Speculation in objects already produced has on the other hand nothing to do with business, and constitutes nothing more or less than a respected form of theft, which the law has done nothing to eliminate. Legislation altogether permits very little to be achieved: it is never constructive. ... We can help the government, but the government cannot help us." Thus Ford is opposed to the middleman and is for the limitation of governmental regulation. He demands production for consumption (good quality and low prices), and rejects it for profit or speculation. He is against the senseless production of a thousand variations by a factory and for the manufacture of fewer models. He is for the rationalization of labor operations that hamper any unnecessary expenditure of effort. He is the enemy of the banks and of banking capital to the extent that they try "to replace work by monetary loans"—and this, he says, is the only reason for his anti-Semitism—and has never availed himself of their assistance. He is against monopolies. He is in favor of every country producing, so far as possible, its own products (and that is no doubt why he has built branch factories in European countries!) He promotes the union of industry and agriculture, and has attached a trade school to his works "so as to enable male youth to be educated and at the same time to receive industrial training on a constructive basis". Finally, Ford is an opponent of war. What, then, do all these things prove? Simply that one can harness a lot of good ideas, even those of socialism and anarchism, in the service of the most inhumane industrial system that has ever been conceived. None of these questions have the least connection with Ford's inner convictions or social awareness; they are all there to advance his undertakings. Nothing else! When workers are hired by the Ford factories no inquiry is made about their previous history, even if they have come from detention or prison; people without arms or legs, or who are blind, etc., are hired. But that is not a consequence of lack of prejudice or humanity, but of the shortage of labor. To the same cause one can ascribe many of Ford's other data about profit-sharing and raises in wages. It is Ford himself who most crassly documents this exclusively utilitarian attitude to the war which impedes his progress: "My opposition to the war is based neither on pacifist principles nor on denial of self-defence. Perhaps our culture is in fact still at a stage which does not permit peaceful settlement of international problems; perhaps they really do have to be fought out. ...What runs counter to all our principles of business is to interrupt the routine of our production except for pressing emergencies."

Everywhere that Ford is not talking about production he reveals amazing obliviousness which, especially as concerns social problems, leads to unconstrained platitudes. Here are just three of his pearls: "Poverty in the world is caused only in the rarest of cases by shortage of goods, but is mainly the result of shortage of money." "Every time that a device is invented that make it possible for one man to do the work of two, the wellbeing of the country is raised, so that new and better employment opportunities are created for the displaced workers." "There are not enough 'rich' and 'poor' to justify such a classification." The exploitation of the product, the profitability of the production is everything to him—but the fate of the people, to the extent he even talks about it, is nothing! "Why have just *one* string to your bow instead of many? ... If pig-breeding goes into recession, why shouldn't the farmer convert his grain into fuel for his tractor? "Ford the man rarely emerges from the background of his enterprises; and it is just this that is entirely characteristic of the Ford works:

nothing there is a personality at work, solely a logically executed principle of manufacturing. Despite the cited contradictions in Ford's data (along with others), I believe it is altogether likely that he is subjectively sincere. His person is consistent with his principles. He worships a religion of tireless work, is contemptuous of "taking-it-easy," and mocks those who believe their lives are "secure"; all connections with culture are cut off, and he confesses a total lack of understanding for history, poetry and art. For him there are only business relations among human beings. "To work hand-in-hand it is not necessary to love one another," Ford says, and I think that his "associates" feel that. He has elevated work itself to the status of a cult—"there is something sacred about a large undertaking that feeds hundreds and thousands of families"—although he acknowledges its effect: "A gigantic corporation is actually too large to be humane. It grows in such a manner that it crushes the personality of the individual. ... The enterprise becomes the important thing." As an American Ford naturally does not forget to assign Christianity (made in the U.S.A. [English in the original]) its proper place and emphasizes that the spirit of America is Christian, and its destiny is to be Christian. Perhaps Ford really believes that his methods will end poverty in the world. But anyone who knows that industrial organization is only *one* of the problems of socialism, that its solution is only an accompanying phenomenon in the realization of a socialist community, will turn away in revulsion from the perspectives that would result from an expansion of Ford's "principles." Let us in conclusion recall the words of Oscar Wilde: "When socialism becomes authoritarian; when it produces governments armed with economic as they now are with political power; when in a word we are in a condition of industrial tyranny: then the final stage of mankind will be worse than the first."

Walter Landauer.

## 111. HENRY FORD

(A rebuttal or complement)

*Der Aufbau* (Zurich) V (22) May 30 1924

The book *My Life and my Work* by Henry Ford was subjected by Walter Landauer to a thoroughgoing criticism in No. 20 of the *Aufbau*. In reading this contribution I could not avoid the impression that its author had not read the book free of prejudice. My assumption was that Mr. Landauer condemned Henry Ford all too skeptically because this voice came from the capitalist camp and therefore was to be heard with the utmost caution. This caution seems then, however, to have been transformed into actual mistrust, and Ford's industrial system is reproached with being "the most inhumane industrial system that has ever been conceived."

I looked in vain in the next issue of *Aufbau* for a rebuttal. None appeared. I regard it, then, as my duty to sketch the impression that Ford's book made on me. Mr. Landauer is in agreement with me at least that it is an interesting book, filled with good ideas, so that it is a pleasure to read the book. But there is something else which is more important: I believe that this man has something to tell the world. Let me therefore fix on some points that are important:

The guiding principle of his production process is *performance.* The driving force of the undertaking is not profit, but performance *vis-à-vis* human society. This is in the first place given expression for the consumer by the delivery of first-class goods. Ford also accepts responsibility for the longest possible durability of his automobiles as the lowest possible price. To that end he employs a host of workers to whom he assures short working hours and high wages. The principle of performance runs like a scarlet thread through his whole life's work. Of real education Ford, for example, says: "What can you do to help the world and to heal it? That is the real test of education."

From the lofty conception of mankind's vocation just cited proceeds a conception of the mission and task of industry that we have never before heard from the class of industrialists. Ford does not regard himself as the proprietor, but as the steward of the capital invested in his factory plants. The profits produced by his enterprises are in the first place plowed back for the benefit of the enterprise that serves his workers as a source of income. In this sense we can certainly agree when he says: "There is something sacred about a large undertaking that feeds hundreds and thousands of families." The profits of the enterprise not only serve the betterment of the workers, but also the consumer. "The task of business activity is to produce for consumption, not for profit or speculation." As an expression of this principle I would mention that at the end of one business year each purchaser of an automobile was given a rebate of $50 because Ford believed that the profit was too high. Ford says in another passage: "Speculation is only a more esteemed form of theft!" and "Energy and machine, gold and goods are useful only insofar as they contribute to freedom in life." He also expresses himself to the effect that ethical principles will govern in all businesses. "One day there will be a universally recognized business ethic."

Even the money institutions are called upon for service. Today they are the masters of industry. Industry is still an object of exploitation for the banks and major capitalists. "We have no room for non-working shareholders." "If I were given the choice between lowering wages and doing away with dividends, I would do away with dividends without hesitation."

Ford also speaks of a connection between agriculture and industry and has in part implemented this idea in his valve factory that was built in the country. Reasons of a sanitary nature are the guide here. He recommends agriculture also for the wealthy classes. "How would it be if we all went on the land in spring and summer, in order to live the healthy life of the farmer for three or four months, instead of spending three or four months in idleness, like the rich at a health resort?"

His observations on poverty are very interesting. He firmly believes that it is possible to eliminate poverty in the world. Ford rejects benevolence that does not have as its ultimate objective to make benevolence unnecessary. "When human feeling impels us to feed the hungry, why does it not produce in us the even more urgent wish to make hunger impossible?" Ford does not only preach this, but has obtained a secure existence for many crippled individuals. He helps them, but not with charity that only undermines their self-confidence. In his factories he displays the proof that cripples placed in the right jobs can produce as much as normal people. Ford gives the statistics on physically handicapped workers employed in his plants. "In a diverse industry there are more jobs for cripples than there are cripples in the world."

Ford knows that his factories are troubled by many shortcomings. He himself says that only a beginning has been made. He knows that his manufacturing methods which exclude as much human labor and thought as possible increases the monotony of factory work. But he tells us on the one hand that the average worker prefers this work, which makes little demand on his ability to think, to a job calling for more intensive work. The worker does not normally ask to get away from his machine. But the statistics given of the turnover in labor personnel, however, do not appear to support this latter assertion. Be that as it may, Ford acknowledges that this is an unsolved problem. "We shall not be a civilized people until the treadmill has been removed from daily work" and "Here a new task beckons: the elimination of monotony."

We do not know Ford personally, and we know his work not from observation, but solely through his book. Therefore our judgment of him cannot be definitive. He is not a man who has solved social problems, but he is a man competent to participate in their discussion; that remains our complete conviction. For this his work and his faith in a better future qualify him. Ford himself says in his conclusion: "Faith is the substance of things we hope for, the proof of the unseen."

Willi Wehrli.

## 112. HENRY FORD AGAIN

*Der Aufbau* (Zurich) V (23) June 6 1924

I would like to make just a few observations on the rejoinder by W. Wehrli in No. 22. First, Wehrli believes that I judge Ford all too skeptically "because this voice came out of the capitalist camp." Well, I think that skepticism could be justified! By chance, however, that was not the case. My reading of this book was stimulated rather by the more or less *approving* opinions of people, mentioned in my article, whose judgment usually coincides with my own, and I hoped that I would find a document of humanity and progress. Despite their very modest success in practice, I acknowledge unbounded admiration for the lofty morality that underlies and underlay the intentions, for example, of Abbe (Carl Zeiss) or Heinrich Freese. There can be no doubt that these men have preserved an enormous sense of responsibility in their enterprises and that they—as, for example Rathenau, too—have sincerely and with the utmost goodwill wrestled with social problems. With Ford it has been otherwise. Reading his book is somewhat less than enjoyable because—as even its translator admits—it is in every way "sloppily" written; but what makes the book offensive is that one repeatedly comes across disgraceful lies, a few of which I have exposed. One can put up with a businessman who conducts his business like everyone else, who is unscrupulous and disregards problems of a social nature, or does not even see them, because he has grown up in a capitalist state and—like most workers themselves—can only think capitalistically. But when someone—like Ford—appears under the pretext of eliminating the class struggle by a new industrial organization then we might—nay, we must demand of him that he adhere punctiliously to the truth in the account he gives of this organization and its methods. But when someone dallies just with the facts as recklessly as Ford, then no one can persuade me to believe in the pretty phrases (work performance, associates, etc.) in which he envelops his account. In my article I deliberately kept close to what Ford says about the development of his works. Since this roast tastes very different from the sauce in which it is served, it seemed to me, and still seems unnecessary to give too many samples of it. Ford's book seems *interesting,* in every way interesting to me, as I had already said before, because it portrays in striking fashion a machine obsessed with production mania, one that does indeed deliver cheaply to its customers, but thereby exhausts all its instrumentalities at a dizzying pace. One might incidentally ask whether and how anybody could imagine the transformation of the whole automobile industry, let alone that of industry altogether, on the basis of Ford's principles?! As one thinks about this question one quickly comes to the conclusion that Ford's grain can germinate only in his present environment. The expansion of his system would probably in the first place result in a further exacerbation of the workers' condition, but beyond that would diminish the temporary benefit to the consumer. Ford's book has indeed much to tell us—let it be a serious warning to many of the ultimate consequences of industrialism heedlessly able to deck itself in all the feathers useful to it (even those of guild socialism or anarchism). Then we might thank Ford, despite everything!

I will point to just one more thing, because it is a very typical Ford argument. Wehrli says: "He (Ford) knows that his manufacturing methods which exclude as much human labor and thought as possible increases the monotony of factory work. But he tells us on the one hand that the average worker prefers this work, which makes little demand on his ability to think, to a job calling for more intensive work." Wehrli copies this passage down without comment! Yet it betrays the whole insanity of Ford's management as repulsively as possible. Of course, all that Ford needs is machines—and let the people wither away miserably, let thought go to the devil, what does that matter, so long as automobile manufacture flourishes!

At the same time as Wehrli's rejoinder, the post has brought me a wonderful little book: Peter Mennicken, *Anti-Ford, or on the Dignity of Mankind* (Die Kuppel, Aachen). Everything that can be said about or against the *man* Ford is in it, as well as anyone could only wish.

Walter Landauer.

## 113. HENRY FORD

(To Mr. Walter Landauer)

*Der Aufbau* (Zurich) V (24) June 13 1924

If the articles about Henry Ford have resulted in *Aufbau* readers themselves looking at Ford's book *My Life and my Work*, and have formed their own opinions of it, then the real purpose of the articles has been achieved.

What forces me to take pen in hand is the interpretation of a concept. I ask you: What do Ford's production methods actually consist of? If we answer that without prejudice we come to the conclusion that Ford really does not have to produce any method of his own. He has simply carried to its logical conclusion a principle that the whole of industry, indeed any single individual follows in more or less perfect fashion in his daily activities, even in domestic matters. It is: In the performance of your obligations eliminate the waste of energy as much as possible. This principle will and *must* be followed in the future, and indeed in a more complete fashion than hitherto. In this regard Ford was only a few steps ahead of other manufacturers. A better social ordering in the future, which we hope for, is needed by men who like Ford possess the talent to organize work (we presuppose: work that is useful to society at large).

Things would be better for us today if we had industrialists who, like Ford, take pride in paying higher wages than other enterprises, and with shorter working hours than the others. More intelligence in plant operations to improve methods of work, instead of reduction in wages and longer hours of work, as we have experienced them in recent times! If in addition these factory managers are imbued with the sense of responsibility to regard consumers and workers not as *objects of exploitation,* but as human beings whom the business is to serve, then that would already be a great step forward. I take Ford to be such a person. I cannot believe that what Ford tells us about his "business ethics" is just bluff, despite the contradictions (which books do not have any?). Even Mr. Landauer says: "Despite the cited (and other) contradictions I believe it is entirely possible that he is subjectively sincere." Contradictions are not necessarily "disgraceful lies."

That Ford is under many illusions about his work as a *social act* is indisputable. Most people believe that their own ideas will be of the greatest benefit to the world. What chiefly struck me as unpleasant in reading Ford's book is the circumstance that Ford sees himself as monarch, as the *autocrat* of his works (an outgrowth of his sincerity!). He tells us flatly that he refused to have dealings with organized trade unions! In this surely lies one important reason for the fluctuations in the work force.

Who knows Ford the man and his works from *his own direct experience?* (perhaps Peter Menniken is such a person?) Let him be heard from in *Aufbau.*

Willi Wehrli.

*Editorial Note:* Ford is certainly a type of manufacturer who has the vision and an awareness of his responsibility, such that many diehard Swiss manufacturers could learn a great deal from him. That his methods represent the elimination of exploitation is hardly subject to discussion. [Treating this as] a solution of the social problem, when he earns so much money, evokes something of ridicule, even if the money finds generous and not ignoble uses. Where does this enormous fortune come from? Surely, either from overcharging the customers or the inadequate payment of the workers, or both. In any case it is exploitation, since one cannot speak of a just wage for Ford's workers in light of such profits.

[Max] Gerber.

## 114. CONSPIRACY AND MURDER IN POST-WAR GERMAN PUBLIC LIFE

*Die Menschheit* (Wiesbaden) XI (22) 8/1/24

There have been political conspiracies at all times. They have everywhere been in the service of retrogression. Still, the history of recent years has distinguished itself above all other periods by the conspiracies which virtually characterize its political confusion. In vain one seeks in the history of other peoples a parallel to a nation decisively defeated in an unprecedentedly savage war of conquest that is still sympathetic to the powers and military leaders responsible for it; that a people which, after all, has to its credit not inconsiderable intellectual achievements and likes to think of itself as a leader in culture, falls prey to the most nefarious demagoguery simply because it is blinkered enough to believe that the economic collapse caused by the war would disappear with a new *regime*; that a nation without passion or even character altogether ignores the judgment passed upon it simply because its own sense of justice has been lost—and finally, that the schemes of reactionary conspirators are not impeded simply because the people themselves do not see the immediate consequences for international relations of their behavior.

E.J. Gumbel has earned much credit by his book *Four Years of Political Murder* by setting out succinctly and clearly the balance sheet of political justice. For years the public have waited for the government to defend itself against the appalling accusations of this indictment. The Ministry of Justice held out the prospect of a Note to the Reichstag on this question. This has now been transmitted to the Reichstag in typescript—after three years—and contrary to custom has not been printed as an official Reichstag document. This Note has been withheld from the public because it is no defense, but a corroboration. Gumbel has now published it with an exceptionally lucid commentary (*The Ministry of Justice Note on "Four Years of Political Murder,"* Malik Press, Berlin). Let us be content here to repeat Gumbel's conclusion: "Word for word the Minister of Justice confirms my assertions, the murders are unreservedly acknowledged, the perpetrators run loose with impunity. To recapitulate its findings again: It is officially confirmed that since 1919 there have been at least 400 political murders in Germany. It is officially confirmed that almost all of them have been committed by right-wing radicals, and it is officially confirmed that the overwhelming number of these murders have remained unpunished." The Note in fact makes no attempt at glossing over, it briefly summarizes the cases, even offers in a very few cases a juridical explanation of the flawed judgments, and only once dwells a little longer on one of the few cases of murder by the left-wing.

This organized political murder would not be possible without widespread, officially favored organizations. In the larger trials it repeatedly emerged that the tracks converged on a few individual places (Ehrhardt, Rossbach etc.). Gumbel has presented in a further book, *Conspirators* (Malik-Press, Berlin) a documented account of these murder centers from which we learn about their actions, their interconnections, their objectives, and as their principal character, their terrifying mindlessness. The empty slogan as a basis for action is surely a trait of German politics of recent years in general, and here it is particularly flagrant.

Anyone who wants to understand the international situation in which Germany was and is placed must read these two books. The mistrust of us by the world will then not only be understood, but shared.

[signed in ink] W. Landauer

## 115. THE PROSECUTION OF DR. J.E. GUMBEL

*Die Menschheit* (Wiesbaden) XI (24) 8/15/24

Among German universities Heidelberg counts as the most progressive. How far in the past that is, the University Senate has now documented by its conduct against its Lecturer Dr. Gumbel. We do not need to waste too many words in these pages about Gumbel's importance for the German pacifist movement. He is one of the few university teachers in Germany (one does not even need the fingers of both hands to count them all) who have spoken up courageously for law and justice, for understanding and reconstruction. It was only recently that we spoke in these columns of his two new very creditable books *The Note* and *Conspirators*).

On the 26th of July, in the Great Hall of the Town Hall, a demonstration—"No more war"—was held at which Professor Balmain of the Sorbonne, Member of Parliament H. Ströbel, and Dr. J.E. Gumbel, as chairman, spoke. In his concluding words Gumbel invited the participants to rise in memory of those fallen in the war and to observe two minutes' silence. In his statement he used the words, with reference to the fallen, "those who, I will not say fell *on the field of dishonor,* but who nonetheless suffered a dreadful death." The assembly accepted this invitation calmly and with dignity, and complied without audible protest—even though reactionary elements were present. Only after the event did the idea come to those to whom Dr. Gumbel, the No-more-war movement and any kind of progress are a thorn in the side, that a noose could be tied about Dr. Gumbel for those words. In a flash a student meeting was summoned which adopted a resolution calling on the university to dismiss Dr. Gumbel. Thereupon (the university disputes a causal connection) the Faculty of Philosophy jointly with the Senate Executive initiated a disciplinary action against Gumbel, and the Ministry of Culture imposed a suspension of Gumbel's teaching activities. Following vehement protest in the Baden parliament, the Ministry has meanwhile withdrawn the suspension. The disciplinary action on revocation of his license to teach continues.

So much for the occurrence and the present state of the matter. We would like to add just a few words of interpretation. For anyone of even slight goodwill the sense of Gumbel's words is clear enough. But even in a formal sense there are really only two assumptions on which the charge can be based. Either the accuser claims that by his words Gumbel has sullied the honor of the war dead, or he finds offense more generally in that the field of battle was characterized as a "field of dishonor." The first assumption is meaningless if one takes Gumbel's words in context. How could Gumbel invite the meeting to rise in solemn memory of those to whom—on this assumption—he offers the worst insult in the same sentence?! One would expect that even the most intolerant reactionaries—among whom the majority of students and university professors must surely be counted—would regard this possibility as beyond dispute! That would leave the second interpretation. And there we clearly understand that to the people supporting the accusation the condemnation of the war and insulting the battlefield seems an unforgivable crime—above all by an academic teacher of youth. But have we really gone so far in Germany as to be able to base a judgment or even an accusation on such emotional opinions? Indeed, we have already once experienced the same—even Nicolai attempted to dishonor war as such.

The university and the Ministry have tried in several publications to draw a parallel with the disciplinary action against Lenard-Mierendorf, and preen themselves on their objectivity and political neutrality. They claim that they are proceeding just as rigorously against the politically left Gumbel as they had then against the reactionary Lenard. In the name of justice! But the university could not compare these two cases without being aware that nothing justifies the comparison. In Lenard's case the issue was that an institute director disobeyed an official order from the Ministry and the university (Memorial Day for Rathenau); Lenard, as an established official, acted in open obstruction of orders of the government against the republic. But Gumbel's "crime", however, is a difference of opinion outside his official duties at which offense could possibly be taken only on logically and humanly impossible grounds, since it was an attack on the emperor's war. But the purpose of this comparison (the *Frankfurter Zeitung* has once again changed sides at the decisive moment, and speaks of "consummate objectivity *vis-à-vis* the nationalist side"!) is too transparent to require explanation. They are looking for a pretext to give the procedure against Dr. Gumbel an appearance of justice in the public eye, and of maintaining even-handed treatment of everybody.

We shall omit passing value judgments. Let the facts speak. All the same—just as the chairman of the local chapter of War Wounded in Heidelberg has done—we express our confidence in Dr. Gumbel, and urge him to continue the struggle to discredit war and the battlefield. May the general public not pass unheeding over these symptomatic events, may each individual consider them important enough to reach a judgment in this case. If so, much perhaps will ensue to help Gumbel's cause, and ours.

Dr. Walter Landauer.

## 116. THE DISCIPLINING OF GUMBEL

*Die Sonntags-Zeitung* (Heilbronn) V (33) 8/17/24

Dr. *Walter Landauer,* a former member of the faculty of *Heidelberg* University, is publishing the following *appeal*:

"The unprecedented events at the University of Heidelberg which have the aim of driving the lecturer Dr. J.E. Gumbel from his teaching position, must engender indignation in every decent thinking individual. Right-wing radical circles, including a number of "liberal" professors, provoked by the cry of "No more war," balk at no means of securing the withdrawal of Gumbel's license to teach. Our best weapon is to affirm to the University our *sympathy for Gumbel* and our outrage at its manner of proceeding against him. We call on everybody, but especially on all those who have been students or members of the University of Heidelberg, to lodge their *protests* by letter or telegram to the University. Anyone who does not speak up for Gumbel supports the reactionaries. Think of how few pacifist university teachers we have, how urgently we need them—and act!

Dr. *Walter Landauer*

Editorial note: The *disciplinary process* against Gumbel, initiated because of a statement made at a "No more war" demonstration, continues, even though the Baden Minister for Education has withdrawn the suspension of Gumbel that was at first imposed.

## 117. ON THE GUMBEL CASE

*Volkszeitung* (Heidelberg) VI (191) 8/19/24

An eerie silence has descended on the *Gumbel* case. No one knows whether this is a good sign, or if there is something ominous in it. The actions of the competent university authorities proceeding, as one is accustomed, at a slow—but surely very "thorough"—pace, have again come to a halt. We hope that the decision is very soon reached. Of great interest is a letter sent to the University Senate by a member of the University about the Gumbel case. It reads:

To the Senate of the University of Heidelberg.

Having been a member of the University of Heidelberg as an assistant until August 1, I feel I have the obligation to protest against procedures that—as it seems to me—must impair the reputation of the University most seriously in the eyes of any unprejudiced thinker within as well as beyond the borders of Germany. The action taken against lecturer Dr. J.E. Gumbel remains outrageous even if seen from the perspective of utterly opposed political views. The only issue here is justice.

From the press releases issued thus far it must be assumed that the suspension of Dr. Gumbel took place in excessive haste and without anything like sufficient deliberation—solely under the pressure of a part of the student body that is immeasurably intolerant, condemning simply from a reactionary political viewpoint. The judgment of the offensive passage in Gumbel's speech—above all by a college of scholars—should have been undertaken in accordance with its meaning and in context—and had it been done in this fashion would have been reduced from the start to the determination as the worst outcome of an unfortunate mode of expression.

I have known lecturer Dr. Gumbel from years of collaboration and am convinced that no member of the faculty of the University surpasses him in moral gravity, in his ethics, and in the purity of his intentions. I believe that the most proper task of the University administration would be decisively to condemn the boorish misconduct of part of the student body; that it was the duty of the University immediately to view matters as anyone of goodwill sees them, and so to clarify the situation.

Now, it seems to me, the task and duty of the University is to remedy as quickly as possible the mistakes it has made. Preserve those people who have maintained their objectivity amid the political confusion and savagery of recent years from feeling ashamed that they have been students and members of Heidelberg University. You should be concerned that the University of

Heidelberg does not present to the world the ignominious spectacle of intellectual slavery or the suppression on any pretext of persons who are a thorn in the side of enemies of the republic.

If Germany wishes once again to fight its way back into esteem among the comity of nations that can occur only by unremitting intellectual honesty, by conscious sense of responsibility for justice, and tolerance towards all the honest intentions of those with clean hands. In the case of Dr. Gumbel the University of Heidelberg faces decisions in accordance with which its moral credit or discredit will be adjudged!

Dr. Walter Landauer.

## 118. A SUMMONS TO MURDER

*Die Sonntags-Zeitung* (Heilbronn) V (34) 8/24/24

"The same segments of the German people do not recognize those foreign commissions and commissioners installed to administer the Dawes Plan as parties acting in fulfillment of legally binding obligations, but see them rather as violators of German independence and sovereignty, *for whose personal safety these segments of the people cannot themselves assume any guarantees.*"

This is an excerpt from a resolution adopted in the last few days in *Weimar* at the *German People's Party Conference.* Adorned with swastika flags the National Theater, supported by a subvention from the German Republic and birthplace of the republican-democratic constitution, was the point from which these programmatic words went forth to party members throughout the country.

We have grown accustomed in recent years that the most shameless provocation and demagoguery take place in the Press and at rallies. We have tolerated it that specific persons have been openly marked for assassination; we have seen how successful this propaganda of the reactionaries has been and will long have to suffer its consequences. But it had so far never happened that a party, as a *party,* so undisguisedly propagandized for murder as the German People's Party now does! Assuring the safety of all persons within the nation's territory is a matter for the nation and for the individual States. Parties have nothing to do with it. What then is it supposed to mean when a party declares that it can assume no responsibility for the safety for certain persons? Can anyone interpret this otherwise than as *an invitation to murder* and its encouragement? We ask, what does the government plan to do to protect the persons under threat and to punish the instigators to murder? Does the Constitutional Court for Protection of the Republic regard this as important enough to occupy itself with the matter?

But the declaration of the German People's Freedom Party contains yet another interesting admission. There can be no doubt that the protocols of the London Conference and the provisions and laws emanating from it were approved and accepted by the German Parliament. What does it mean when the German People's Freedom Party declares that it will not recognize those persons charged with the execution of the laws enacted by the German Parliament as "acting in fulfillment of *legally binding* obligations"? If this open proclamation of disobedience of the laws of the land is not *high treason,* then what would one understand by that? We address this question expressly to the guardians of the law! And we ask the German Parliament what it has to say when one of its parties—and today no longer one of the small ones—sets itself above the enactments of the people's representatives and calls upon its members to deny them recognition?

For the first time in years the German government has taken a serious step toward reconciliation and the reconstruction of the international community. Never more than now was the cry more urgent: *Caveant consules*![1]

Walter Landauer

[119-121 (Henry Ford);122-123 (US farmers'cooperatives omitted]

## 124. FREEDOM OF THE INDIVIDUAL IN POPPER'S *GUARANTEED SUBSISTENCE FOR ALL*

*Mitteilungen des Vereins "Allgemeine Nährpflicht"* (Frankfurt), May 1925

"...one thing is certain. I have *worked* my whole life, and I have undertaken tasks for altruistic purposes; I have worked for myself, for others, and quite intentionally for the advancement of mankind."

"And I am, like any other worker, "worthy of my hire", that is to say: I am satisfied."

This final conclusion by Popper in his *Autobiography* (1st ed., p. 100) made a deep impression on me. He grew up under the most pitiful conditions; the

---

[1] [The whole phrase is: *Caveant consules ne quid detrimenti respublica capiat!* - Let the consuls beware that no harm come to the republic!]

misery of his own existence became for him the symbol of the misery experienced by others; the brutality of the struggle for life fed the unquenchable source of his human goodness, of the all-encompassing readiness to help that impelled his actions. His love of problems of a scientific-physical nature took second place in favor of his greater love of human society. His ever alert feeling of unconditional responsibility and unlimited sense of belonging filled his life and work with intellectual honesty to a truly rare degree. That is how Popper strode through life, his impressions causing him unspeakable suffering. But that suffering was turned in his charitable heart into love, in his clear analytical understanding into a brilliant design for a new social structure. Material need and poverty characterized Popper's life from beginning to end. Nevertheless: "I am satisfied!" The unconscionable conditions under which people are forced to live today impelled him to abandon the problems which most captivated his mind, because those conditions daily caused him new and infinite pain. Nevertheless: "I am satisfied!" This satisfaction leads us into the most intimate understanding of Popper the man. This satisfaction is a reflection of the most beautiful qualities of his human personality: goodness and modesty. Popper gazed deeply into the essence of human nature, he saw acutely the good and bad sides of his generation—his *Fantasies of a Realist* are witness to this. His writings on the reshaping of society are grounded on the most immediate necessities of existence, but at their root are basic ethical demands. In the first place Popper wanted to assure each individual's bare existence and only then say: now you can also be a bit more human. He struggled with himself to attain a life of goodness and modesty—that was the struggle in his heart; he fought with others to secure the basic rights of the individual—that was the struggle in his mind. He was certain that neither his heart nor his mind would give up the battle—therein lay his satisfaction. Moved by this deep admiration for Popper's human qualities as, indeed, for his work, I add some critical or complementary remarks.

"*An individual is something that does not wish to end its existence, but wants to exist in its own way*"—this fundamental thesis of Popper (*The Individual and the Evaluation of Human Lives*) is the key to all his social prescriptions. From this point of departure he demands for every individual "the right to live" (minimal subsistence), and rejected "the duty to die" (military service). But if the individual's right to guaranteed subsistence was to be preserved at all, Popper found himself at once compelled to limit the right "to exist in his own way." "An unbounded respect for human life and for the physical inviolability of every individual—*insofar as it does not imperil the lives of others*—and piety, are posed as the fundamental conditions of an ethical culture..." (*Autobiography*, p. 88). Elsewhere Popper says that his program actualizes "*Freedom* in economic matters, *insofar as is ethically permissible,* that is, reconcilable with the safeguarding of all economic practices." (*Fundamentals for a New Political Constitution*, p. 37 [Carl Reissner, Dresden, 1905]). And in a similar vein this limitation on individual freedom for ethical reasons recurs frequently. The principal difficulties that arise here are clearly recognized by Popper, who expressly designates them "basic sensibilities," and continues: "Whether, then, it is *this* sentiment or the opposite [i.e. Popper's own] assuring economic existence through coercion, instead of allowing life to perish in order to preserve freedom—whether this sentiment or the other will prevail cannot be predicted. The struggle between them and their proponents will settle the matter. Both of them are justified since neither is refutable: one can say: "I'd rather starve to death"; the other : "I'd rather let others starve to death before I support entry into or submit to a compulsory organization, or even have them represent me." (*Guaranteed Subsistence For All*, 1st ed., pp. 365/6). I do not accept the arguments of those who support the current economic system for egotistical interest. But I believe that what emerges from this situation with the utmost clarity is that we must explore every possibility "of so organizing one's own life—without doing harm to others"—that *the least possible coercion* is exercised upon the individual and his inclinations in order to remain true to the principles laid down by Popper himself. "Therefore, no one should be subjected to a *tyranny of idealists,* because they are not agreed among themselves, and even if they were of one mind, none of them has the right to force others beneath the yoke of their lifestyles or perspectives on the world" (*Autobiography*, pp 89/90).

What then is the route that Popper envisions for the realization of his guaranteed subsistence program? He differentiates between "fundamental" and "secondary" human needs and says that a general plebiscite of the people must decide on the necessity of securing the satisfaction of the fundamental needs of all people, while the secondary needs could be governed by current political principles (parliamentarianism). In both cases, however, majority decisions would apply to establish the new economic system. But Constantin Frantz (Germany and Federalism) has brilliantly demonstrated the proof how often enough unmistakable minorities have asserted their will by means of "majority decisions." I will only give a hint of this. Let us just assume that this majority decision is true to the facts, i.e., there is a minority—perhaps even a very large one—that is obliged to submit. What consequences then emerge for its implementation? Inescapably it could come about only by means of a central governmental administration, a centralism that is all the more rigorous the larger the minority is, against whose will it must assert itself. And Popper comes to this conclusion too. He expresses this in several ways. I cite just two examples. Take his definition of the State: "The State is in many respects an instrument of coercion equipped with the greatest physical power; this institution imposes itself on all native-born or resident inhabitants of a geographically delineated area. And all these citizens live together, under normal circumstances, in a stable dynamic equilibrium between many more or less tacitly or vocally bearable or unbearable demands" (*War, Conscription and the Constitution*, p. 230). And the other citation, which answers the question about the jurisprudential principle of work assignments in the "Labor Corps," reads as follows: "To this the simple answer is that the 'Labor Corps' is a specifically militarily organized body. Without even remotely postulating on that account anything like the harsh discipline of our armed forces, one still has to view the orders of the relevant administrative authorities as a coercive condition under which there can be no talk of a legal principle in the exercise of power. To be *completely* free and independent, and *at the same time* to be safeguarded against need is quite impossible in our latitudes" (*Fundamentals for a New Political Constitution*, p. 40). Completely free, surely not, but if one accepts Popper's basic ethical demands one must strive to guarantee the *maximum possible* freedom, and one would do that even more if one views as false Popper's thesis in "We ourselves are the State" (*Guaranteed Subsistence For All*, p. 244). Concessions made to centralized power organizations are very promptly paid for. Even Popper had to pay for them. Fritz Wittels, who has most profoundly steeped himself in Popper's brilliant train of thought, and who was the first, to my knowledge, to substitute the word "community" for "State," reports in his wonderful book *The Elimination of Need* (p. 69): "I brought to the attention of my mentor the question whether the learned professions would not be harmed by the guaranteed subsistence program." Terse and harsh, resentful and ominous, Popper is said to have replied: "I cannot help the scholar!" "But," adds Wittels, "Popper did not wish it said of him that he was ever harsh, resentful, or ominous," and so he softens the passage in Wittels' book into the following text: "I cannot help the scholar. He, too, must serve in the subsistence army. We shall find ways of assuring his uninterrupted education." But Popper found himself betrayed in another passage into a much harsher measure as a result of the centralizing principle needed to execute his program.[1] He speaks of the ways of getting dangerous and unpleasant tasks done, and continues: "anyone who seeks to withdraw would, in the most extreme case, be denied his minimum subsistence, and he would not be able to get it in any manner other than through the subsistence authorities, since all such commodities would be under an embargo" (A debate with Socialism and the Socialists, p. 6). Would that then mean that the guaranteed subsistence State reserves the right to allow its

[1] I note here that many similar passages are to be found in his book *War, Conscription and the Constitution*. This one book of Popper's, to which I am in principle opposed, seems to me to be the outcome of the intellectual situation described above. I do not dwell on it further, because the book is not of direct significance for the mandated subsistence program.

members to go hungry or to starve to death? No, we cannot agree with that! No, we want to proclaim the unconditional and inalienable right of the individual to minimum subsistence! We find an interesting parallel in the excellent, but little noticed book by Wilhelm A. Wilhelm, *Economic Democracy in the Future* (Anzengruber, Vienna) in which we read (p. 236): "The law must prescribe under what conditions a citizen able to work can acquire and exercise membership in a public economic community of workers, *and under what preconditions he forfeits these rights.*" And then? Even here this disquieting question remains unanswered.

I'll be brief. The guaranteed subsistence program is an ideal solution, not of the social question, but of the *most pressing* material-social question. Its implementation by a centralized administrative system would seriously jeopardize the basic rights of the individual, and would deprive him to a substantial degree of the ethical basis which can exist only in the realm of freedom. Popper himself assumes that it will take many decades before his program is ripe for realization. Why should we not employ this time to found his ideas within the structure of a decentralized community (unions, councils, guilds, cooperatives)? We have that obligation as soon as we recognize that this is the only way to guarantee the most far-reaching integrity of the individual. How much can be achieved, without waiting for new laws, the (by no means ideal as yet) cooperative movement has shown. Would it not be possible that propagandizing the guaranteed subsistence program along these lines might achieve the goal even sooner? Let us advocate moral autonomy, associations of independent individualities, the governance of social conscience![1]

Let us build bridges to a new world, the world of Popper, in which the rights of man are founded on his "having-been-born" and "being-alive" (Nökel)! Let us find the surest way to preserve the rights and duties of the individual and of the community!—Above all, let us reconcile the two!

Storrs, Conn. (U.S.A.), November 1924.

*Walter Landauer.*

[1] Defined by Karl Nökel [erratum: Nötzel is meant, also in the paragraph below] in his pioneering book *Introduction to Socialism without Dogma*, as the triad of "certainty of a duty," "certainty of a failure of duty," "certainty of a never completely possible fulfillment of duty."

## 125. THE CASE OF GUMBEL
## A Contribution to Understanding the German Mentality

*Der Aufbau* (Lucerne) VI (27) 7/3/25

In all the years since the collapse of the German military and the takeover of the bankrupt estate of the Hohenzollern kingdom by the Republic, the German republicans have more and more come to terms with living under a system of secret courts. Outside of dwindling circles without influence, the norm has been lost that sets limits on how far one can go without incurring the contempt of even the most ruthless power politician and most hardboiled capitalist magnate. The policies of political exclusion under the empire have been replaced by a policy of moral exclusion under the Republic.

The most substantial foundation for an analysis of the spiritual brutalization of Germany: the history and statistics of national murder, has been furnished us by E.J. Gumbel. Still, it will not be solely on account of his personal merits—not easy to overestimate—that we shall speak of the "Gumbel case," but rather because this case illustrates in a particularly drastic manner the growing unscrupulousness of the nationalists against something else—against a new Germany.

In January of this year a trial of Gumbel for high treason on account of speeches he had made before the "French League for Human Rights" was terminated because there were no facts to support the charge. In April three treason trials against Gumbel were terminated for the same reason. One of them was based on an article on "Rossbach," which contained only material long since published in the democratic press; the second was directed against Gumbel's book *Conspirators*, and the third finally was directed against an essay published in *Menschheit*: "Is there a 'black' army?" Heidelberg University, in which Dr. Gumbel is a lecturer, commenced a disciplinary action against Gumbel in July last year, and applied to the Baden Ministry of Education for the revocation of his license to teach. The reason was that Gumbel had used the expression "field of dishonor" at a "No More War" rally. The immediate provisional revocation by the Ministry of his right to lecture was promptly withdrawn, and the University decided to suspend its action until the abovementioned high treason and treason trials were over. One can hardly go wrong in interpreting this wait-and-see posture of the University as conveying the hope of the hearing committee that one at least of these trials would lead to a conviction, and so furnish a convenient basis for the expulsion of Gumbel from the academic faculty. And it does not require much imagination to envisage their disappointment at the negative outcome of the trials. But these people do not even sense the need to conceal their personal feelings of hatred; their fury at their failure can be read in every word of the judgment that has now been published. The Academic Committee of Inquiry declares that "from the outset it had believed that an evaluation of the details was possible only within the ambit of the *over-all personality.*" The impression conveyed by Dr. Gumbel's personality was that of a "decidedly demagogic nature." We had better leave aside the extent to which, beyond that, he also possesses those talents that are the precondition of fruitful scholarly work. A University thus does not consider the scholarly work of a university teacher to be significant enough to be considered in judging his "over-all personality"; or while it does not regard itself as competent to evaluate scholarly activities, it is surely able to address political ones instead. It is not uninteresting to recall the extremely mild handling at the same University of the Lenard case (he was guilty of obstruction in opposing the governmental order for a day to commemorate Rathenau), in which they relied on the towering scholarly standing of Lenard. It is clear that Gumbel's judges had in some fashion formed a picture of him as a scholar, and it is probable that they knew that he could call on the brilliant expert opinions of Einstein, among others. But how could they have done that without weakening their judgment of his over-all personality, and how then could they have continued: "this much is certain, that *in his political activity not even the slightest trace of scholarly qualities can be detected!"*

"Intellectual level at a serious nadir," "in substantive and stylistic respects there is a total lack of objectivity," "distinctly facile in intellect," "volatile change of viewpoint," "distortion and mutilation of the objective account of matters," "a virtually elementary lack of tact," the lack of all "objectivity and intellectual dignity," a failure to take into account the need "to evidence self-restraint in face of the hostility he has provoked (cf. the appearance of Helmuth von Gerlach he sponsored, and of a Frenchman speaking in the French language [in the *French* language!—W.L.] at a session of the Peace Society the day after January 18 1925); "in the area of moral judgment—a lack of feeling for all more refined nuances," "political fanaticism"—thus run the value judgments hurled against Gumbel. And it is on the basis of these judgments—and they form the whole findings of the process; not an atom of *fact,* not even an approach to an objective statement—that the faculty declares "that Dr. Gumbel's membership in it appears *distinctly unwelcome* to them"!

This is not the place to analyze the specific value judgments, or to demonstrate their subjective and nationalistic basis; and before the forum of *Aufbau* readers is as little necessary as the proof that in the judgment of Gumbel's "over-all personality" what has been ignored in a tendentious manner—consciously perhaps?—is the most important aspect of the evaluation of a university teacher, his activity in research and teaching. Let us underscore just a few things here. As its central charge, the faculty accuses Gumbel of "an elementary lack of tact." To an almost unbelievable degree this brings the hearing committee itself into contempt. The real outcome of the process cannot be other than the declaration: Dr. Gumbel is and remains (if only temporarily and unfortunately) *our colleague.* And this same faculty which says: "The Faculty sees in his indifference to *corporate solidarity* (My italics—W.L.) a danger to its unitary functioning" in a consciously subjective fashion smears the reputation of a colleague in the daily press! "A virtually elementary lack of tact"!

The local chapter in Heidelberg of the German Peace Society correctly points to the pharisaical hypocrisy of the charge of "volatile change of viewpoint," since the faculty repeatedly applied to the Education Ministry for the revocation of his license to teach, and now in the name of the *principle of academic freedom* pretends to gloss over the suspension.

In August last year, in a letter to the Senate of Heidelberg University which also appeared in the daily press, I took a position on the process against Gumbel. I was at that time reproached as being subjective and premature in my statements. This referred to the following points: 1. I had characterized the suspension of Gumbel as "undertaken in an overhasty fashion and without anything like sufficient investigation." The withdrawal of the suspension a few days later *proved* that this action was played out prematurely and without sufficient previous investigation. 2. I had criticized the University because they had not represented the affair as "anyone of goodwill" would have seen it. As to this I could support my position by the following facts: 1. At the "No-more-war" rally no protest had been voiced against Gumbel, and the insult to national honor was concocted only after the event. The meeting stood at Gumbel's invitation in full understanding to honor those who fell on the field of dishonor. 2. The "League of War-Wounded" issued the express declaration that its members in no way felt insulted by Gumbel's utterance. 3. The metropolitan press emphasized the "moral gravity" of the speaker and the "deep spiritual emotion" of the meeting. With the judgment of the faculty that has now been published, it is surely unnecessary to look for more evidence to prove their lack of goodwill. And, finally, I had said—admittedly not very gently—in apprehensive anticipation of further events: "You should be concerned that the University of Heidelberg does not present to the world the ignominious spectacle of intellectual slavery or the suppression on any pretext of persons who are a thorn in the side of enemies of the republic." My worst fears have been surpassed by the results of the process. At least as regards my opinion of the way the University court has handled the matter. The expulsion of Gumbel that one feared has indeed not taken place (it is typical that they do not simply say that there are no grounds for the revocation of the license to teach, but that the judgment is based on an interpretation of the applicable regulations in the 1805 ordinances, such that "the Gumbel case cannot be subsumed under that rule"). I cannot prove that their aim was *deliberate,* but anyone who knows German conditions knows that the effect of this publication by the governing circles of the University is: 1. To induce the students to boycott Gumbel; 2. To make Gumbel's situation impossible not only in Heidelberg but in all other German universities; 3. If possible to "settle" Gumbel (or "do him in").

Finally, let us cite from the individual opinion of one of the members of the disciplinary tribunal the following remarkable passage: "He believes in one cause, pacifism, and in his commitment to it. He stands passionately and filled with hatred in opposition to everything that seems to him to be violence, nationalism, or an inclination to future war. When this idealism is at issue he has the courage, not just the courage of a civilian, to say what he thinks, but also the courage to place his life at risk." We congratulate Dr. Gumbel on this praise and for the confirmation by the Philosophy Faculty of Heidelberg University that speaking as a university teacher in behalf of pacifism and fighting hatred and violence entails placing one's life at risk. And we hope that despite everything the possibility still remains beyond his activity as a researcher to work for the discrediting of war and of the field of honor. May he also be unwelcome to some in so doing.

Walter Landauer, Storrs (Conn.), U.S.A.

[126 (The lives of others); 127-130 (American commentary); 131 (Crime astiny) omitted]

## APPENDIX: GENEALOGICAL TABLE *

**GRANDPARENTS**

| | |
|---|---|
| **Hermann Landauer**, merchant | m. **Rosa, née Neuburger** |
| b. 9/21/1837 in Buttenhausen | b. 10/28/1845 in Buchau/Federsee |
| d. 2/9/1900 in Karlsruhe | d. 12/17/1932 in Karlsruhe |
| m. 6/28/1864 in Buttenhausen | Parents: David Neuburger and Jette, née Einstein |

**THEIR CHILDREN**:

| | |
|---|---|
| 1. **Friedrich**, District Judge | m. **Charlotte, née Ziegler** |
| b. 2/3/1866 in Karlsruhe | b. 11/13/1866 in Heidelberg |
| d. 5/5/1901 in Mannheim | d. 2/2/1943 in Heidelberg |
| m. 7/3/1894 in Mannheim | |
| | |
| 2. **Felix**, merchant | m. **Johanna, née Würzburger** |
| Owner of his father's business | b. 7/25/1873 in Heilbronn |
| b. 4/29/1867 in Karlsruhe | d. Sept.1966 in New York |
| d. 1/8/1939 in Karlsruhe | |
| m. 4/21/1898 in Heilbronn | |
| | |
| 3. **Gustav**, author | m. 1.**Margarethe, née Leuschner** |
| b. 4/7/1870 in Karlsruhe | b. 5/11/1872 in Berlin |
| d. 5/2/1919 in Munich [murdered] | d. 1/14/08 in Berlin |
| m. 1. 1894 | 2. **Hedwig, née Lachmann** |
| 2. 1903 | b. 8/29/1865 in Stolp/Pomerania |
| | d. 2/21/1918 in Krumbach |

**PARENTS**

| | |
|---|---|
| **Hugo Landauer**, merchant | m. **Charlotte, née Ziegler** |
| b. 7/10/1868 in Buttenhausen | b. 11/13/1866 in Heidelberg |
| d. c.1933 in Meersburg/Dasendorf | d. 2/2/43 in Heidelberg |
| m. 9/4/03 in Überlingen/Bodensee | |

* A more detailed genealogy is deposited in the Landauer Archive, Dodd Research Center, University of Connecticut Libraries.

**CHILDREN**

1. **Hermann** (from Charlotte's first marriage to Friedrich Landauer)
   b. 6/1/1895 in Philippsburg
   d. after 1963 in New York
      Emigrated to Chile c. 1938; emigrated to USA c. 1946; married
2. **Walter** (from Charlotte's first marriage to Friedrich Landauer)
   b. 7/15/1896 in Mannheim
   d. 2/19/1978 in London
      Emigrated to USA 1924
      m. 1. Anne 1935; 2. Naomi 1940; 3. Annie 1954
      4. Elly 1964, née Bernstein, widow of Kiewe
3. **Heinz**
   b. 11/15/1904 in Überlingen
   d. after 1933 (mentally retarded, institutionalized, euthanized)
4. **Erich Otto**
   b. 12/19/1905 in Augsburg
   d. c.1982 in Puerto Rico
      Was director of the Health Service in Shanghai, China.
      m. c.1930 to Yuen; later to Stella (2 sons, 1 daughter; all musicians)
      A grandson, Timothy Landauer, is a cellist in USA.
5. **Fritz**
   b. March 1908 in Pforzheim (?)
   d. after 1960 . Drug addicted in China, returned to Germany in 1945
      at the request of his daughter, Eva; was institutionalized
6. **Liesel**
   b. 5/8/1910
   d. 1929 in a traffic accident

**PARENTS**

**Gustav Landauer**, author
b. 4/7/1870 in Karlsruhe
d. 5/2/1919 in Munich [murdered]
Studied in Heidelberg, Strasbourg, Berlin

m. **Hedwig, née Lachmann**
b. 8/29/1865 in Stolp/Pomerania
d. 2/21/1918 in Krumbach

**CHILDREN**

1. **Charlotte Clara** (from his first mariage to Margarethe Leuschner)
   b. 1/5/1894 in Berlin
   d. 8/13/1927 in Karlsruhe
   m. 3/14/1922 in Karlsruhe to Dr. Max Kronstein, chemist
   b. 10/7/1895 in Basel
   d. 1944 in USA
   Daughter: **Marianne Rahel Kronstein**, b. 5/13/1923 in Karlsruhe
   (1/5/1939: Dr.Max and Marianne Kronstein emigrated to USA)
2. **Anna** (from his first marriage to Margarethe Leuschner)
   b. 5/19/1896 in Berlin
   d. 8/14/1898 in Berlin
3. **Gudula**
   b. August 1902 in Berlin
   d. 1946 in New York (in a traffic accident)
4. **Brigitte**
   b. 4/10/1906 in Berlin
   d. October 1985 in Philadelphia
   m. 1. **Paul Nicholaievitch Peschkowsky** (Nichols), physician; d. 1944
   Children: **Mike Nichols**, film director, b. 11/6/1931 in Berlin
   **Robert Nichols**, neurosurgeon, b. 9/14/1935 in Berlin emigrated to USA in 1939
   m. 2. **Dr. Franz X. Hausberger**, 1950, in Philadelphia

## BIBLIOGRAPHY

Butler, Rohan D'O. 1942. *The Roots of National Socialism*. E.R.Dutton & Co. N.Y.

Clark, H. and L. Pierro. 1994. "Walter Landauer: The Man and his Work." Issues and Reviews in Teratology 7:217-283.

DuBois-Reymond, Emil. 1912. *Reden*, ed. By Estelle DuBois-Reymond, Leipzig. Cited by Charles E. McClelland in "The Wise Man's Burden" in *Essays on Culture and Society in Modern Germany*, edited by Gary Stark and Bede Karl Lackner. Texas A. and M. Press. 1982.

Fishman, Sterling. 1960. *Prophets, Poets and Priests: A Study of the Men and Ideas that Made the Munich Revolution of 1918/1919*. University Microfilms. Ann Arbor, MI.

Foerster, Friedrich. 1920. *Mein Kampf gegen militaristisches und nationalistisches Deutschland* (My Struggle Against Militaristic and Nationalistic Germany). Verlag Friede durch Recht, Stuttgart.

Fromkin, David. 1999. *The Way of the World*. Knopf. N.Y.

Gasman, Daniel. 1971. *The Scientific Origins of National Socialism: Social Darwinism in Ernst Haeckel and the German Monist League*. MacDonald. London.

Geiss, Imanuel. 1970. "The Outbreak of the First World War and German War Aims," pp. 315-326 in *The European Past*, Vol. II, edited by Clough, Shepard B., Gay, Peter, Warner, Charles K., Cammett, John M. Macmillan, N.Y.

Gerber, Max; Matthieu, Jean; Ragaz, Clara and Leonhard; Staudinger, Dora. 1922. *A New Socialist Program*. W. Troesch, Olten.

Gumbel, E.J. 1924. *Vier Jahre politischer Mord* (Four Years of Political Murder). Verlag der neuen Gesellschaft. Berlin.

Jansen, Christian, 1991. Emil Julius Gumbel: *Portrait eines Zivilisten.* Wunderhorn Verlag, Heidielberg.

____________. 1924. *Verschwoerer* (Conspirators). Vienna.

Kaufmann, Walter H. 1973. *Monarchism in the Weimar Republic*. Octagon Books. New York.

Kershaw, Ian. 1998. *Hitler: 1889-1936*. W.W.Norton. N.Y.

Landauer, Gustav 1919. *For Socialism*, 2nd Ed. Cassirer Verlag, Berlin. Translation by David J. Parent. Introduction by Russell Berman and Tim Luke. Telos Press. St. Louis. 1978.

Lunn, Eugene. 1973. *Gustav Landauer. Prophet of Community*. University of California Press.

Lutz, Ralph H. (Editor) 1932. *Fall of the German Empire, 1914-1918* (Hoover War Library Publications, No. 1, pp 74-78). Stanford University Press.

Maurer, Charles B. 1971. *Call to Revolution.* Wayne State University Press.

Mitchell, Allan. 1965. *The Revolution in Bavaria, 1918: The Eisner Regime and the Soviet Republic.* Princeton University Press.

Mosse, George. 1961. *The Culture of Western Europe.* Rand McNally & Co. Chicago.

Nicolai, G.F. 1918. *The Biology of War* (Translated from the German by Constance A. Grande and Julian Grande) The Century Co., N.Y.

Pinson, Koppel S. 1954. *Modern Germany.* MacMillan Co. N.Y.

Popper (or Popper-Lynkeus) Josef. 1878. (reprinted, 1924) *Das Recht zu leben und die Pflicht zu sterben* (The Right to Live and the Duty to Die). R. Loewit Verlag. Vienna, Leipzig.

__________. 1905. *Fundament eines neuen Staatsrechts* (Fundamentals for a new Political Constitution). Verlag Carl Reissner. Dresden.

__________. 1912. *Die allgemeine Nährpflicht als Lösung der sozialen Frage* (Guaranteed Subsistence for all as a Solution of the Social Question). Verlag Carl Reissner. Dresden.

__________. 1917. *Selbstbiographie* (Autobiography) Verlag Unesma. Leipzig.

__________. 1920. *Das Individuum und die Bewertung menschlicher Existenzen* (The Individual and Evaluation of Human Lives) Verlag Carl Reissner. Dresden.

Preuss, Hugo. 1919. *Die Staatsverfassung von 1919* (Germany's Republican State Constitution)
Neuer Staat Press. Berlin. (The text of the Constitution is reprinted in an appendix.)

Scanlon, T.M. 1998. *What We Owe To Each Other.* Harvard University Press. Cambridge.

Stern, Fritz. 1961. *The Politics of Cultural Despair.* University of California Press. Berkeley.

Stroud, Robert. 1944. *Digest of the Diseases of Birds.* Privately published. A copy now housed in Landauer Archives, Dodd Center, Univ. of Connecticut.

Wachtel, Henry L. 1955. *Security for All and Free Enterprise.* Philosophical Library. N.Y.

Wittels, Fritz. 1995. *Freud and the Child Woman: The Memoirs of Fritz Wittels.* Edited by Edward Timm. Yale University Press. New Haven.

Varia by and about Gustav Landauer, Collected by Walter Landauer, deposited in Landauer Collection of Dodd Archives, University of Connecticut Libraries.

## PERSONS NAMED IN THE PAPERS

We have endeavored to include all the individuals mentioned in the papers. In a few cases we were unable to trace them in the many biographical and bibliographical reference works consulted. Numbers in bold type refer to the numbered papers, not the pages. A few organizations referred to in the text are briefly described below.

**Orgesch** (Organisation Escherich). Originally a "Protective League" in Bavaria. Dissolved by the Allies (Kaufman) and revived by the German National People's Party throughout Germany; antirepublican and monarchistic (see Guenther Axhauser, 1921, Organisation Escherich, Berlin).

**Stahlhelm.** A "legal organization" without political agenda, the "Prussian Orgesch." Antirepublican, made up of discharged soldiers, especially former officers of the army. After 1923, three divisions: battle veterans, former officers with less than six months' service, and youth (Kaufman).

**Thule Society.** Antirepublican organization, strong in Bavaria (not listed, however, by Kaufman among "militant leagues and secret organizations" in the early years of the Republic).

**Vorwärts.** Social Democratic Weekly founded by Wilhelm Liebknecht in 1876. Discontinued February 28, 1933. Reappeared as Neue Vorwärts in Prague, 1933-1937, and Paris, 1938-40. In 1955, started anew as Vorwärts under new editorship.

Gorky, Maxim (1868-1936), **71, 90**. Writer, political theorist.

Grabowsky, A. (1880-1965), **20**. Politician, jurist.

Graefe, Albrecht (1868-1933), **30**. German People's Freedom Party; Reichstag 1919-29.

Grelling, Richard (1853-1929), **4, 26**. Pacifist; author of "J'Accuse"; co-founder of German Peace Society.

Gumbel, E. J. (1891-1966), **115-117**. Statistician, pacifist.

Hegel, G. W. F. (1770-1831), **11, 74**. Philosopher.

Heine, Heinrich (1797-1856), **33, 75, 90**. Poet

Helfferich, Dr. Karl (1877-1924), **15**. Economist; Nationalist politician.

Hergt, Oskar (1869-1967), **15**. Finance, Nationalist, Reichstag 1930-32.

Herwegh, Georg (1817-1875), **13**. Revolution of 1848; prolific writer.

Herzen, Alexander (1812-1870), **12**. Russian writer, lived in London, Paris, Brussels.

Herzog, Wilhelm (1884-1960), **13**. Leftist writer.

Hesse, Hermann (1877-1962), **28, 43**. Nobel Prize for Literature, 1946.

Hilferding, Rudolf (1877-1941), **30, 31**. Socialist politician, economist, pacifist; Finance Minister 1923-29.

Hindenburg, Paul (1847-1934), **12**. Military leader; President 1925-33.

Hippius, Zinaida (1869-1945), **71**. Russian writer.

Hobbes, Thomas (1588-1679), **41**. Philosopher, monarchist.

Hoffman, Johannes (1890-1967), **6**. Bavarian politician.

Hölderlin, Friedrich (1770-1843), **7**. Poet.

Hubacher, Hermann ( ? - ? ), **31**.

Hugenberg, Alfred (1865-1951), **15**. Nationalist politician, industrialist.

Jaurès, Jean (1859-1914), **74**. French Socialist, assassinated 1919.

Juliusberger, Otto ( ? - ? ), **2**.

Kant, Immanuel (1724-1804), **12**. Philosopher; Professor of Logic, Königsberg.

Kardorff, Siegfried von (1873-1945), **15**. Nationalist politician.

Keller, Gottfried (1819-1890), **101**. Swiss novelist, dramatist, poet.

Kerschensteiner, Georg (1854-1932), **9**. Nationalist; Professor, Munich.

Kierkegaard, Søren (1813-1855), **26, 40**. Philosopher, religion.

Klabund (1861-1928), **2**. Pseudonym of Alfred Hensche, distinguished writer.

Kolshorn, ( ? - ? ), **7**.

Kraemer, Hermann ( ? - ? ), **9**.

Kropotkin, Peter (1842-1921), **21, 41, 90**. Russian anarchist, thinker, Mutual Aid.

Landauer, Gustav (1870-1919), **11, 20**. Socialist, writer, Bavarian revolution; murdered 1919.

Ledebour, Georg (1850-1946), **12, 31**. Socialist politician; leftist.